Soldier's Study Guide

*A Guide to Prepare for Promotion Boards
and Advancement*

7th Edition

CSM Walter J. Jackson
USA (Ret.)

STACKPOLE
BOOKS

0 11557 01189 0

Library of Congress Cataloging-in-Publication Data

Jackson, Walter J., 1944–
  Soldier's study guide : a guide to prepare for promotion boards and advancement / CSM Walter J. Jackson, USA (Ret.). — 7th edition.
    pages cm
  ISBN 978-0-8117-1189-0
  1. United States. Army—Examinations. 2. United States. Army—Promotions. 3. United States. Army—Vocational guidance. I. Title.
  U408.5.J33 2013
  355.0076—dc23

                                                                    2013024034

*This guide is dedicated to my father, James H. Jackson,
who taught me values when I was young;
to the senior noncommissioned officers
who taught me my chosen profession;
and to the youth of today's Army
who will be the noncommissioned officers of tomorrow.*

"I'd like to have two armies: one for display with lovely guns, tanks, little soldiers, staffs, distinguished and doddering generals, and dear little regimental officers who would be deeply concerned over their general's bowel movements or their colonel's piles; an army that would be shown for a modest fee on every fairground in the country.

The other would be the real one, composed entirely of young enthusiasts in camouflage uniforms, who would not be put on display but from whom impossible efforts would be demanded and to whom all sorts of tricks would be taught. That's the army in which I should like to fight."

—Jean Larte'guy

"War is an ugly thing, but not the ugliest of things. The decayed state of moral and patriotic feelings which thinks nothing is worth war is much worse. The individual who has nothing for which he is willing to fight, whose only concern is for his own personal safety, is a miserable creature who has no chance of being free unless made and kept so by the exertions of better men than himself."

—Author Unknown

Contents

Introduction

This guide has been prepared to provide some direction and guidance for the enlisted soldiers of the United States Army. The information comes from military publications, established general policies, and my own military experience of more than twenty years' service. Although this guide is based on Army regulations, field manuals, training manuals, and general policy, it is my own work and does not necessarily represent official government policy.

This book is more than just a study guide for promotion boards. It contains the ground rules for being a better soldier and the information to be a better supervisor and NCO. It is a guide to get the most from your time in the military. There are other resources for learning. Manuals, online or web-based study guides, and hands-on training are three of them. This book compliments those assets and is usable when a computer or electrical power is not available. It is a tool to help you advance. It is a tool you may find useful during your entire military career. If you committed three questions a day to memory, within a year your military knowledge would be phenomenal.

Every soldier has the potential and the opportunity to advance to the highest enlisted grades or to become a commissioned or warrant officer. Much of the opportunity offered in the Army is wasted, however, because many soldiers do not work to their full potential. They do just enough to get by and then hope that promotion and advancement will come.

Many soldiers believe that advancement is automatic; it is not. Each soldier must compete for the next available slot. Modern equipment, advanced technology, new programs, and an ever-changing force structure mean only the best reach the top pay grades. Although there are many soldiers in today's Army, there can be only so many sergeants, first sergeants, and sergeants major.

Everyone desires success. It is more fulfilling to spend twenty or more years in the military and retire as a sergeant major than to spend twenty years and retire as a staff sergeant. Even if a soldier plans to spend only three or four years in the service, he or she will find it more reward-ing to be discharged as a sergeant than as a private or a specialist.

It is not that difficult to get promoted and attain positions of respon-sibility if you are willing to work a little harder to do well. Things of value have a cost. If you want to advance and have a successful Army career, you need to set priorities and then follow them. It is up to you to decide how far you will advance. Now is the time to start.

1

The Secrets for Advancement

Despite the title, the information in this chapter is no secret. Career counselors, first sergeants, and unit commanders have expressed these ideas to individuals and small groups for several years. These recommendations based on Army regulations, policies, force structure and general practices are presented here in simple terms. The real secret is in believing the study of these ideas can lead you to discover how to apply them to your own life.

Set a Goal for Yourself

Many people in our society make lists prior to shopping and planning their vacations. In so doing, they both accomplish their goals and save time and money. If we plan simple tasks such as going to the store or taking a vacation, doesn't it make sense to plan a goal as complicated as a career? Setting a goal means more than just thinking or saying that you would like to be the Sergeant Major of the Army. It means taking stock of your present strengths and weaknesses and then outlining the steps that will get you to your goal. Some people mentally note what they need prior to going shopping; others inventory their cupboard, make a list, break the list down into like items, figure out the store with the cheapest cost, gather coupons, and then hit the road. Both methods get the job done. But in following the first way you are more likely to forget something, buy things you don't need, and pay more than necessary. It is your life. No one else is going to set your goals and your career for you.

Too many soldiers drift through the service. They accept their situations and surrender control of their lives to the "system." Because the Army is made up of people, and people make mistakes, the system has problems. If you let someone else control your life, your only recourse is to live with the decisions that are made for you.

If you don't have a goal, you will find that it will be difficult to take advantage of opportunities; you will drift among the masses with little hope for advancement. Focusing on a goal allows you to make correct decisions without having to ponder over them. Motivated by a goal, you will be more likely to recognize opportunities. When you display an attitude for success, your actions will positively affect your career. The good things in life have a cost; most of your decisions will require trade-offs with other appealing avenues in your life. Fortunately, the price you must pay for success is not high.

Use your career counselor, platoon sergeant, first sergeant, and sergeant major to help you advance, to provide you with counseling, or to help you make wise decisions. Although officers can help, too, senior noncommissioned officers (NCOs) are the most knowledgeable about enlisted soldiers and the programs that pertain to them.

Live by the Standards
In the civilian world, people obey local, county, state, and federal laws. In the military, soldiers live by standards, regulations, policies, and orders. The only difference is that the military more directly controls the individual.

Military regulations, policies, and orders are similar to the local, county, state, and federal laws by which civilians live. Standards are the daily scales against which soldiers are measured; not making the standard is failure, exceeding the standard is the road to success.

People often judge others with very little information. Those who can have the greatest effect on your military career are, unfortunately, those who see you the least. Your first sergeant, unit commander, sergeant major, and battalion commander take in a limited impression of you when they see you in formation, while they are walking through the unit motor pool, or when they inspect your barracks. Although you may be the hardest-working soldier in your company, if you need a haircut every time that the sergeant major sees you, he will form a negative opinion of you. And when you appear before a promotion board or a soldier-of-the-month board, he will probably be on it.

Successful leaders surround themselves with dependable people. Why? Because good people make their job easier, lessen their stress, and help them to look good. Remember: Take care of the boss by meeting and exceeding standards, and the boss will take care of you. In the military, all those in higher ranks than yours can affect your advancement. Creating a positive image is simple: look, walk, talk, and act like a soldier to convince everyone that you are a good one. Creating a positive impression from the start is much easier than changing a bad one. If you comply with the established standards, and exceed them when you can, you will positively impress your superiors.

The chain of command seems to spend 90 percent of its time with problem soldiers and the other 10 percent with the outstanding ones. If this is true, then those who merely get by (average soldiers) earn little recognition. To advance, be in that small percentage of above-average performers.

Demonstrate Your Potential

The enlisted system is designed to promote the best-qualified soldiers: the key word is "best-qualified." Being average or below average will not advance you in today's Army. Below-average performance will invite a Qualitative Management Program (QMP) screening and a bar to reenlistment. Average performance may take you to the rank of staff sergeant and twenty years of active service. A bit more effort can land you in the top enlisted grades. Dedication and working toward a goal can make you a sergeant major, a warrant officer, or a commissioned officer.

The Army schools individuals who demonstrate that they are ready to perform at the next higher pay grade. Demonstrate that readiness. Volunteer for less than enjoyable tasks, take charge in the absence of a supervisor should the situation arise, appear before local boards, excel at school, and "be all you can be."

Appear Before Boards

Within the Army structure, boards are the standard procedure for the selection of the best-qualified individuals for schools, awards, and promotions. It is very important, therefore, to prepare for board appearances. Promotion boards and other local boards allow you to demonstrate your potential. The ability to present yourself before a soldier-of-the-month board, school selection board, or promotion board as a knowledgeable, confident soldier marks you as someone ready for advancement. Your leaders remember the achievers.

Many soldiers shy away from appearing before local boards until forced to go before a promotion board. This is the wrong attitude. Not making first place is disappointing, but you still have gained something positive. Appearing before a board expands your knowledge and experience, and spotlights you to those who can help you get ahead.

Is there a cost for demonstrating your promotion readiness before a board? Yes, the price is the time and energy spent in preparation. But it is a small price in comparison with the good that can result from a successful board appearance. Too many soldiers don't want to invest the time and energy. They appear before boards unprepared—not only hurting themselves but also embarrassing their immediate supervisors and their units. Board members can recognize when a soldier has expended the effort to do well.

Serve your best interests by taking the needed time to present the best image possible. You don't have to prepare alone; remember that your NCO support channel is there to help. Seek advice from your platoon sergeant or first sergeant. In chapter 3 you will learn not only how to be recommended for a board appearance but also how to make it work in your favor.

Keep Your Tools Bright and Shiny

I'm not taking about physical tools; I'm talking about skills and knowledge. Learning things is like putting tools in a tool box. You never know when you will need them, and they are of little use if they are broken or rusty.

Search Out Schools

Attending schools is essential to a career; their inconvenience at times is one of the costs of success and rapid promotions. Never turn down a school unless your attendance would create a severe hardship for you or your family. Many soldiers either avoid going to school or attend but put forth less than 100 percent effort. Rather than study in the evening, these soldiers party, watch TV, or play video games. Their academic reports indicate that they did just enough to get by.

Yet, attending a school to increase your knowledge is a prime way to demonstrate your potential to your unit and the Army. Those students who try harder usually return to their units with reports that recommend them for more schooling and positions with greater responsibility. The price is small compared with the gains: more knowledge, recognition by your chain of command, administrative promotion points, and positive additions to your personnel file. The points will help with high cutoff scores in local promotions. And the positive image in your file will impress Department of the Army (DA) boards in their choice of the top three grades.

Reach for Positions of Responsibility
Welcome greater responsibility, which comes with position as well as rank. A master sergeant and a first sergeant, for example, both have E-8 pay grades. The master sergeant may have more time-in-service and time-in-grade than the first sergeant. But the first sergeant is in charge because he has the authority and responsibility that is delegated from the company commander.

Many soldiers shirk responsibility because they don't want to make or live with decisions. They desire the money and the advantages that come with promotions, but they don't want the problems that accompany responsibility. It is not realistic to expect to be promoted without being willing to shoulder responsibilities.

Volunteer to Do Unpleasant Tasks
And then do them well. When your supervisors realize they can count on you, more work will come your way. Don't complain. That recognition will also bring the support, the breaks, and the promotions that will carry you up the ladder to more rank, higher pay, and positions of even greater responsibility.

Perform Daily Tasks to the Best of Your Ability
Each day is an opportunity to show your potential to your chain of command through your appearance and performance. Complete the simplest of jobs to the best of your ability, whether alone or with others. When you pay attention to details, your performance will stand out. Take the extra step to clean the trash can as well as empty it. If told to GI (clean) the floor, get into the corners and then clean the baseboards as well. Always try to do your best—no commander can ask more.

Above all, act smart. For example, don't speed or run stop signs if you are late for formation. Not only will you still be late, but you also may have to explain a ticket if you are stopped by the military police. Don't drink and drive. Avoid drugs. Respect members of the opposite sex. Shun contact with racial bigots. Be responsible. Think before you act. If you make a mistake, admit it. Learn from it, and do your best not to repeat it.

Remember: Your daily performance includes your off-duty time, too. Unlike civilian jobs, where your nonwork behavior doesn't affect your employer, in the military you're a soldier twenty-four hours a day, seven days a week. Your conduct constantly reflects on you, your unit, and the

Army. Soldiers with promising careers have lost them because they couldn't discipline themselves after normal duty hours.

Many times during Article 15 proceedings I heard an accused soldier's chain of command explain to the battalion commander that here was a "good soldier" who had "just made a mistake."

Somewhere in the statement would be an admission that the soldier knew better. Good comments aside, the accused usually received a stiff punishment. It is not that commanders and old sergeants major believe that mistakes don't happen. Honest mistakes are expected and can be forgiven. But misconduct is the result of a bad decision. And good soldiers with leadership ability don't willfully make bad decisions.

When a soldier knowingly violates a policy or regulation, the chain has to question the soldier's ability to perform in responsible positions. The same applies to accidents. Accidents are usually the result of a bad decision, horseplay, or carelessness with detail, three things that good soldiers and leaders try to avoid. As the Army troop strength drops, it will become more difficult to recover from bad decisions (especially misconduct). Drug use, alcohol abuse, spouse abuse, sexual harassment, racial bigotry, fraternization, misuse of government property, willful disobedience, or a felony arrest can, will, and should terminate a career. Think twice before you willfully and knowingly violate policy, regulations, or civilian law.

Improve Yourself

Even if you have a master's degree, can "max" every event in the physical training (PT) test, are an expert on the rifle range, can run for days, and know your job backward and forward, you can still better yourself. Life's real winners compete against themselves. The happiest people are those who are always seeking out new things to learn and do.

Everyone has room to improve. Every soldier in your pay grade desires that same promotion. The farther you climb the ladder, the more you must compete for the next higher rung. You alone control how competitive you are going to be.

The following are some ways you can improve your chances for promotion, no matter what your present pay grade is or to what grade you aspire. On local boards they affect the promotion points you need for the next higher grade. On Department of the Army (DA) boards, they are discriminators that may determine your selection for a school, assignment, or promotion. To succeed you need to excel.

Civilian Education

College is becoming ever more important to a soldier's success and promotion. A two- or four-year degree is a positive promotion discriminator for the top senior grades. College courses taken during your junior-grade years give you promotion points. And, the classes themselves can only help advancing soldiers with the challenges of leading and training soldiers. New distance-learning tools, such as *eArmyU* and other college education programs available through the education centers, make it easy for the soldier to get college credits, promotion points, and usable knowledge. The college courses taken or degrees you obtain while in the military will enhance your civilian prospects when you leave the military.

Army Correspondence Courses

The Army Correspondence Course Program offers military education courses and should be part of your game plan for advancement in technical skill and responsibility, and for furthering your professional development. A course consists of one or more subcourses to support a specific professional field, skill qualification development, MOS-related tasks, and primary leadership subjects. Courses are designed for the average student to read the material and complete all exercises and the examination. Like college credits, correspondence courses earn promotion points—one point for every five subcourse hours.

Physical Condition

Everyone has the ability to be more fit. Some people hate to run; others don't like to do push-ups. Success, however, comes from doing what is necessary. To improve, work harder on those areas that cause your lower points. Just passing is average; aim to "max" the text. You also need to work to improve your latest score. If you are not getting better, you are starting to go in reverse. Improving your physical condition will also improve your ability to recover from illness or injuries. It improves your stamina, reduces stress, and allows you to recover more quickly from hard work or long hours.

Height and Weight Standards

Look like a soldier. No matter what you believe about the connection between your ability to do the job and being overweight, the military has standards. To meet them, exercise more, eat less, and get below the maximum weight for your height. Work to present a trim "fit to fight" appearance. This will increase your chances for promotion, as well as ensure better health, less stress, and maybe a longer life.

Weapons Qualification

Improvement in this area may be difficult, but it is still within your power to do. Using a weapon demands concentration. The average person cannot take a weapon from the arms room and automatically qualify on the range as an expert. Practice is needed, but not necessarily practice with live ammunition. Use the dime/washer or shadow box techniques as explained in FM 3-22.9, M16A1 and M16A2 Rifle Marksmanship, to help with sight alignment and trigger control. Also practice on the simulators that are available on most qualifying ranges. As others joke and loaf while waiting to qualify, spend your time practicing with the available equipment.

Testing and Training

The DA-directed individual soldier testing programs of the past are gone, but training and testing go on in the Army. The major issue with soldier testing and training is that their efficiency varies from unit to unit and leader to leader. Published standards do not guarantee all units receive the same level of training or that all soldiers reach the same level of proficiency. To excel, a soldier must seek self-development and practice necessary skills.

FM 6-22, Army Leadership, states that Army leaders are trained by "three equally important pillars: institutional training, operational assignments, and self-development." The one most fully controlled by the individual is self-development. All soldiers are expected to meet three standards: proficiency in their jobs (MOS), knowledge of basic combat skills, and development of their leadership traits. These goals are easy to achieve by making a schedule.

Devote as little as thirty minutes a day or a few hours on the weekend to master required skills. Do this and you will do well on any test. Even more importantly, if sent to a battlefield, you'll have the skills and knowledge to do your job and survive.

Armed Services Vocational Aptitude Battery (ASVAB) and General Technical (GT) Scores

Your ASVAB scores were used to determine your career field in the Army. A low GT score is an obvious negative discriminator for assignments, schools, and promotion. Army Education center programs can help you raise your scores and become more competitive. You can also improve your reading and writing—skills that become more important with each promotion—by taking courses. Paperwork and complicated training manuals face today's leaders. NCOs are expected to document counseling, to write lesson plans, and to correspond with individuals outside their unit. To reach the top enlisted grades, it is crucial to increase your education level, both military and civilian.

Learn About and Use the System

Having served for more than twenty years in the Army, I'm convinced that the military tries to take care of the soldier and that the soldier's needs and desires are placed right up there with the needs of the Army. The Army will not know your individual needs or desires, however, unless you reveal them.

During my first twenty years in the army, I attended an average of one school a year. The opportunities to attend these schools didn't seek me. I had to go after them. If I was denied my request for school, it never hurt me. I just asked again.

Promotions in the Army are governed by regulations that spell out how to make them happen. If you are not promoted, read the regulation that covers promotions to find out the requirements. Then you will be able to talk with your first sergeant with some idea of what should be happening and where you stand. To determine if you qualify for a particular school, read the regulation or see your career counselor. Ignore the rumors. Instead of blindly accepting all you are told, read the regulations. It is easier to do this today than ever before; the regulations are just a couple mouse clicks away. Question the system, and use the system to better both yourself and the Army.

Using the system means more than just knowing it and using it to achieve your desired goals. You need to ensure that the soldiers in the system (clerks, supervisors, leaders) make the system function correctly. Clerks update records, but you must check your records each year to see if the information in them is up-to-date. The clerk might not know that you attended a school, received an award, or qualified with your weapon on the rifle range.

Never stay in a dead-end job. Some military occupational specialties (MOSs) are too small for promotions to come quickly. Don't be afraid to go where the promotions are. Fear of the unknown has kept many a good soldier from the ladder of success. Cautious soldiers might say that they don't mind the slow promotions in fields that they enjoy. After years of no promotions, many of them change their minds.

Remember: No one can take away the skills that you have learned. Mastering new jobs makes you more valuable to the Army. Each additional MOS opens up an avenue for advancement.

Volunteer for tough jobs. Good soldiers should spend an assignment or two learning new skills. Time spent in Special Forces, Rangers, drill sergeant duty, recruiting and career counseling, and reserve component duty teaches skills you will never learn anywhere else. These assignments also are positive discriminators when a DA board looks at your records for an assignment or promotion.

Caution: Spend only four to six years in special assignments and then rejoin a regular Army unit. I advise this for many reasons:

- Senior NCO positions in these areas are scarce, except in Special Operations.
- If you excel in these areas, you will outshine soldiers in regular units.
- Adapting to the leadership challenges in a regular Army unit is difficult after spending more than six years in special units.
- Marriages suffer because of time-consuming assignments, and divorce may result.
- Pack animals we are not. Jumping from aircraft and humping heavy rucks have caused many old soldiers to live with a bad back, knees, or ankles. Such intense physical demands are better made on young bodies.

There is a term called "ticket punching." Ticket punching is obtaining training and assignments just to get promoted. Obtaining jump wings or a Ranger Tab and never serving in a unit where you can use these skills is a waste of Army resources. Ticket punching is not what I am recommending. I am recommending these assignments to better both the soldier and the Army. And while I do caution soldiers about extended time in a "special assignment," there are those soldiers who make excellent Special Forces soldiers. Those that can should stay for a number of reasons. The first is that only a few who try, make it. The second is that it takes four to six years just to train and season a Special Forces soldier. The third is that there are too few who really have that warrior spirit and dedication it takes to be a Special Forces soldier.

Develop Leadership Traits

Expertise and the desire for promotion will not get you very far if you lack the traits necessary to be a good leader. Ambition, decency, and the work ethic single out a person in the Army. How does a good leader behave?

- A good leader updates the knowledge needed to perform, seeks new skills and challenges, and shares knowledge with his or her comrades.
- A good leader arrives first and leaves last. He is willing to do and has done in the past the disagreeable tasks he must ask of his soldiers. A leader is the first to tackle a dangerous situation and departs only when his or her soldiers are safely out of harm's way. Long ago, when I was a platoon sergeant in Germany, my squad was often detailed to the battalion support task of cleaning the motor pool wash rack at the end of the day. At times I helped shovel out the trash and dirt. Every time that I did, a squad member would generously take the shovel from me. The other platoon sergeants didn't think I should be there, much less pick up a broom or shovel. But I gained from the brief time spent with those squads at the wash rack. Would you rather have a boss who willingly helps or one who delegates dirty tasks while he parties? What kind of leader will you be?
- A good leader puts mission first, troops second, and himself last. He or she takes care of soldiers' needs before his own. He's honest with soldiers and divides workloads evenly.
- A good leader is a positive example to those who learn by watching him. Your example can be good or bad. If you feel guilty, ashamed, or secretive, you are doing something wrong.
- A good leader is a planner. He or she considers the effect his or her actions will have before acting. The carpenter's old rule "measure twice and cut once" is great advice to follow.
- A good leader accepts the responsibility for the results of his actions and the actions of his or her soldiers. Too often I've overheard NCOs refusing to take responsibility for the actions of their units. In effect, the leader blames the group for whatever went wrong and relieves himself of any failure. If a leader plans the mission well (seeing to training, equipment, supplies, support, motivation, supervision, etc.), the failures and mistakes will be minimal. But, even if the careful leader is still chewed out, he will suck it up, take the reprimand, and do all that can be done to prevent another one. A good leader accepts the blame for mistakes just as he accepts the praise that comes with success.
- A good leader delegates. Delegation accomplishes two things. The first is easing the leader's load. No one can do it all. Details will be overlooked and mistakes will be made. The second result of delegation is the training of future leaders. Being trusted to

perform provides a sense of accomplishment in soldiers and builds confidence between them and their leaders. As a by-product, the leader learns who will be of help in a crisis.

- A good leader follows orders as easily as he or she gives them.
- A good leader is proactive. Reactive leaders do what needs to be done at the moment. They dawdle over tasks that subordinates could easily do, drink coffee, or make small talk with friends. Proactive leaders are always checking, planning, and solving problems before they become major headaches. When the boss says, "Look into this," they usually have already cleaned it up or have a ready answer.

The NCO is the backbone of the Army, the officer corps is the head, and the soldier is the muscle that gets things done. The muscle is no good without the bone structure to attach to and pull against. The head, useless without support, needs something to control.

Be All You Can Be

The Army once advertised "Be all you can be." And you can, if you put forth the effort. Would you be wearing Army Blue if you didn't have the ability to advance and be successful? We come into the Army with skills we have developed and skills that need developing, as well as shortcomings that we must deal with. Some struggle with the standards, and others must work harder to achieve. But if you have made it past basic training, you have the ability and the opportunity to rise to the top.

The military is an unusual employer because it spells out in writing what your job is, how to do your job, and what you need to do to get promoted. Armed with this knowledge, the choice of what to do with your skills and opportunity is up to you.

Learn the Basics

The Army changes, and publications change with the adoption of new weapons and techniques and the needs of the military. But some things do not change: the basic skills and conduct that make a soldier a good soldier and an NCO an excellent one. The basics that make a good soldier or NCO today are the same ones applied in World War II, Korea, and Vietnam. Honesty, candor, integrity, knowing your job, seeking self-improvement, and taking care of fellow soldiers will carry you a long way.

2

Duties and Responsibilities

Consider the fact that you are in a unit where all have received the same basic training, have completed an MOS course, and are experiencing the same leadership. What makes a platoon sergeant, first sergeant, or commander select one soldier over another for mentoring and advancement? Evidently, in addition to basic job knowledge, they are seeking the following qualities in a soldier: accomplishing tasks well, doing more than what is expected, following rules, working well in a team, and accepting responsibility.

A major problem caused by the continual assignment and reassignment of soldiers is that not all are trained the same. Some receive more or better training than others. Many soldiers don't do as well as they could because they don't fully understand the scope of their duties and responsibilities.

Responsibility is being accountable for your actions. *Duties* are tasks that must be done because of a specific responsibility. There are many daily tasks based on areas of responsibility in training, soldier welfare, discipline, maintenance, accountability, and execution of the mission.

A soldier must be responsible. His or her fellow soldiers and commanders rely on that responsibility. Without it, the Army cannot function at its fullest capabilities. Commanders and sergeants cannot watch every soldier twenty-four hours a day. Those soldiers who demonstrate that they are responsible for their own actions and ready to assume the responsibility for the actions of others will be the ones to get promoted.

As a memory refresher for older soldiers and as a guide for new soldiers, the following task list is provided. It gives some idea of the many tasks that soldiers and NCOs should be doing each day. It is by no means all-inclusive.

Task	Person Responsible		
	First Sergeant	**Platoon Sergeant**	**Squad Leader**
Spot Check Daily			
Motor pool (equipment status)	X	X	X
Arms room (weapons' cleanliness)	X	X	
NBC room (equipment status)	X	X	
Supply room (operations, cleanliness)	X		
Commo shop (equipment status)	X	X	
Barracks (cleanliness, maintenance)	X	X	X
Inspect areas for proper police	X	X	X
Check vehicles for trash and PMCS (preventative maintenance checks and services)	X	X	X
Check for SAFETY all the time	X	X	X
Check for physical security violations	X	X	X
Soldiers' appearance	X	X	X
Soldiers' welfare	X	X	X
Training			
Ensure that personnel who require certification attend the training	X	X	X
Ensure that soldiers attend scheduled classes	X	X	X
Ensure that all personnel take part in physical training and meet standards	X	X	X
Create an environment of discipline	X	X	X
Develop future leaders	X	X	X
Train as you will fight	X	X	X

Person Responsible

Task	First Sergeant	Platoon Sergeant	Squad Leader
Maintenance			
Know the commander's guidance; assist in accomplishing maintenance goals	X	X	X
Ensure that services are performed on time		X	X
Check on deadlined equipment		X	X
At the close of the day, check the motor pool line for oil spills, trash, unsecured vehicles, and vehicles parked on line	X	X	X
Ensure that PMCSs are being performed when required		X	X
Accountability			
Ensure that tools and equipment are marked	X	X	X
Inspect tools and equipment for presence and serviceability		X	X
Conduct monthly inventories for squad tools and equipment		X	X
Inspect soldiers' CTA-50 items after each field problem and/or each quarter		X	X

Good soldiers have been doing and checking on the tasks listed here for a long time. Good NCOs will accomplish many of the listed tasks each day. Exceptional NCOs will accomplish most of them in a given day. Good soldiers who want to become NCOs or officers will ensure that when an NCO or officer checks on them, their "stuff" will be straight. They will have performed to the best of their abilities.

3

Boards

Promotion eligibility and procedures governing promotions and reductions are described in AR 600-8-19, Enlisted Promotions and Reductions.

Local Promotion Boards

Local promotion boards use a point system to determine which soldiers should be placed on a promotion roster. Points granted in a wide range of areas support promotion under the "whole person" concept. The gains and losses in each pay grade, and grade strengths with budget limitations mandated by Congress, affect the point cutoff levels for each individual MOS. Local selection boards recommend soldiers in the grades of E-4 and E-5 for advancement to the next higher grade. Individual soldiers can do much to influence their chances for recommendation and advancement when appearing before their local boards.

A total of 800 points is possible when you are being evaluated for promotion to E-5 and E-6. One hundred and fifty of these points are awarded by the board. To appear before a board, you must be recommended by a supervisor and the unit chain of command. The unit commander must evaluate your potential before forwarding the necessary paperwork to the promotion authority. In most cases, the promotion authority is the battalion commander. Areas your commander will consider are results of tests, weapons qualification, level of responsibility, efficiency, maturity, time-in-service, and time-in-grade. The biggest factor will be your commander's opinion of your ability to perform at the next higher pay grade. Don't wait a week or two before your local board to start working for promotion. Almost everything a soldier does, day after day, week after week, affects his or her chances for promotion.

Promotion points are calculated using a DA Form 3355, Promotion Point Worksheet. The maximum number of points you may be awarded is broken down into six areas:

1. Military training	100 points
2. Duty performance (commander's points)	150 points
3. Awards and decorations	100 points
4. Military education	200 points
5. Civilian education	100 points
6. Board points	150 points
	Total 800 points

Board points are given by averaging together the points granted by each voting member of the board. DA Form 3356, Board Member Appraisal Worksheet, defines the point spread that may be granted in six different areas. These are the areas a board member will be concerned with and evaluating:

1. Personal appearance, bearing, self-confidence	25 points
2. Oral expression and conversational skills	25 points
3. Knowledge of world affairs	25 points
4. Awareness of military programs	25 points
5. Knowledge of basic soldiering	25 points
6. Soldier's attitude and potential	25 points
	Total 150 points

Study the subject areas of DA Form 3355, Promotion Point Worksheet, and DA Form 3356, Board Member Appraisal Worksheet, (shown on the following pages) so that you can become familiar with the areas in which you can improve your point posture. Use the graded areas as a guide to set goals and to develop a self-improvement program. It is also very important that you review your records before your board appearance.

Preparations are necessary in other areas. First, your records should be as complete and as accurate as you and the clerk can make them. Review your DA Form 2-A, 2-1, and Military Personnel Records Jacket at least fourteen days before the promotion board. Make sure that all of the information about you is current and accurate. Another area is your uniform and appearance. Your uniform should be neatly pressed, set up, and fit you as indicated in AR 670-1, Wear and Appearance of the Army

PROMOTION POINT WORKSHEET

For use of this form, see AR 600-8-19; the proponent agency is DCSPER

1. TYPE	2. DATE *(YYYYMMDD)*
☐ a. Initial	
☐ b. Total Reevaluation	

DATA REQUIRED BY THE PRIVACY ACT OF 1974

AUTHORITY:	Title 5 USC, Section 301.
PRINCIPAL PURPOSE:	To determine eligibility for promotion.
ROUTINE USES:	Reviewed to determine promotion eligibility and validity of points granted.
DISCLOSURE:	The furnishing of fraudulent information may result in denial of promotion.

3. NAME	4. SSN	5. RECOMMENDED GRADE
6. ORGANIZATION	7. PMOS	

SECTION A - RECOMMENDATION

1. MILITARY TRAINING *(Maximum 100 Points)*

a. LATEST APFT DATE *(YYYYMMDD)*	b. SCORES				c. POINTS AWARDED
	PUSH-UPS	SIT-UPS	RUN	TOTAL	

d. LATEST WEAPONS QUALIFICATION DATE *(YYYYMMDD)*	e. DA FORM USED:	f. TOTAL HITS	g. POINTS AWARDED

h. TOTAL POINTS AWARDED ⟶

2. DUTY PERFORMANCE EVALUATION *(Maximum 150 Points Award 1-30 Points For Each Category)*

CATEGORY	POINTS AWARDED
a. COMPETENCE: Proficient, Knowledgeable, Communicates Effectively	
b. MILITARY BEARING: Role Model, Appearance, Confidence	
c. LEADERSHIP: Motivates Soldiers, Sets Standards, Mission, Concern	
d. TRAINING: Individual and Team, Shares Knowledge and Experience, Teaching	
e. RESPONSIBILITY AND ACCOUNTABILITY: Equipment, Facilities, Safety, Conservation	
f. TOTAL POINTS AWARDED ⟶	

I certify that the above APFT and weapons qualification scores shown have been extracted from appropriate records and the latest valid scores are in accordance with Army Training Regulations and Army Field Manuals.

3. SIGNATURE OF COMMANDER	4. TYPED OR PRINTED NAME AND GRADE	5. DATE *(YYYYMMDD)*

SECTION B - ADMINISTRATIVE POINTS

1. AWARDS, DECORATIONS AND ACHIEVEMENTS *(Maximum 100 Points. List all awards individually. Include award number (i.e. 3rd OLC) and the order number.)*

TOTAL POINTS AWARDED ⟶

DA FORM 3355, MAY 2000	PREVIOUS EDITIONS ARE OBSOLETE	Page 1 of 2	USAPA V1.00

NAME						SSN		

SECTION B - ADMINISTRATIVE POINTS *(Continued)*

2. MILITARY EDUCATION *(Maximum 200 Points. List all military education.)*

TOTAL POINTS AWARDED ————————————————————➤

3. CIVILIAN EDUCATION *(Maximum 100 Points. List all civilian education.)*

TOTAL POINTS AWARDED ————————————————————➤

I certify that the above administrative points shown have been accurately extracted from appropriate records and that the promotion points indicated are correct.

4. TYPED OR PRINTED NAME OF RESPONSIBLE OFFICIAL	5. DATE *(YYYYMMDD)*	6. SIGNATURE OF RECOMMENDED INDIVIDUAL *(Required)*	7. DATE *(YYYYMMDD)*

SECTION C - TOTALS

Only whole numbers will be used in awarding promotion points for all sections (drop fractions). Only initial and total reevaluations require submission of DA From 3355. Administrative reevaluations and adjustments are submitted on DA Form 4187 and annotated in the Eval/Adj column.

1. POINTS GRANTED

ITEM	INITIAL *(Date)*	EVAL/ADJ *(Date)*	EVAL/ADJ *(Date)*	EVAL/ADJ *(Date)*	EVAL/ADJ *(Date)*	EVAL/ADJ *(Date)*
a. TOTAL PERFORMANCE EVALUATION AND MILITARY TRAINING POINTS - SECTION A *(Maximum 250 points)*						
b. TOTAL ADMINISTRATIVE POINTS - SECTION B *(Maximum 400 points)*						
c. TOTAL BOARD POINTS *(Maximum 150 points)*						
d. TOTAL PROMOTION POINTS *(Maximum 800 points)*						
2. INITIALS OF RESPONSIBLE PSB OFFICIAL						

SECTION D - CERTIFICATION

I certify that the above total points shown have been accurately extracted from appropriate records and promotion list points indicated are correct.

1. RECOMMENDED BY BOARD ☐ YES ☐ NO		2. ATTAINED MINIMUM POINTS ☐ YES ☐ NO	
3. TYPED OR PRINTED NAME AND SIGNATURE OF BOARD RECORDER		4. GRADE	5. DATE *(YYYYMMDD)*

I certify that the soldier has been recommended for promotion by a valid promotion board.

6. TYPED OR PRINTED NAME OF PROMOTION AUTHORITY	7. SIGNATURE	8. DATE PROCEEDINGS WERE APPROVED *(YYYYMMDD)*

Counseling statement: I have been counseled on my promotion status and deficiencies. *(Use only when recommendation is disapproved, when a soldier is not selected by a board, or when a soldier cannot be added to the recommended list due to not attaining the minimum required points).*

9. SIGNATURE OF SOLDIER	10. DATE *(YYYYMMDD)*	11. TYPED OR PRINTED NAME AND SIGNATURE OF COUNSELOR

BOARD MEMBER APPRAISAL WORKSHEET

For use of this form, see AR 600-8-19; the proponent agency is DCSPER.

1. NAME	2. RECOMMENDED GRADE	3. RECOMMENDED MOS

4. Board Interview and Evaluation and Points Awarded

AREAS OF EVALUATION	AVERAGE (1-7 Points)	ABOVE AVERAGE (8-13 Points)	EXCELLENT (14-19 Points)	OUTSTANDING (20-25 Points)	TOTAL POINTS
a. Personal Appearance, Bearing, and Self-Confidence					
b. Oral Expression and Conversational Skills					
c. Knowledge of World Affairs					
d. Awareness of Military Programs					
e. Knowledge of Basic Soldiering *(Soldier's Manual) (See note.)*					
f. Soldier's Attitude *(includes leadership and potential for promotion, trends in performance, etc)*.					
				g. TOTAL POINTS AWARDED	

NOTE: Questions concerning the knowledge of basic soldiering will be tailored to include land navigation, survival, night operations, inclement weather operations, adverse environment, and terrain.

5. REMARKS

6. ☐ I DO   ☐ I DO NOT   Recommend the Soldier for Promotion

7. SIGNATURE OF BOARD MEMBER	8. RANK	9. DATE *(YYYYMMDD)*

DA FORM 3356, MAY 2000   PREVIOUS EDITIONS ARE OBSOLETE   USAPA V1.00

Uniforms and Insignia. Your overall appearance is a major factor in gaining points from the board. Minor things such as unbuttoned pockets, misplaced brass or awards, loose threads, or a tight uniform blouse could cost you needed points. Set up your uniform days before the board and have it inspected by your platoon sergeant and first sergeant.

Practice answering questions. Learn to speak up. Give a positive impression to the board members. You are going to be nervous, and the board expects it. Board members won't expect you to answer all the questions correctly, but they will expect you to demonstrate a fair amount of military knowledge, to be prepared, and to present a confident, take-charge attitude.

The actual conduct of a board will differ from unit to unit and board president to board president. The contents of each board, however, are fairly constant. The practice questions provided in this guide cover most areas you will encounter. A local board normally consists of three to five board members, a recorder, and a board president (usually a command sergeant major). If the president is a voting member, he will normally vote only to break a tie.

Some board presidents will ask, "Why should this board recommend you for promotion?" Answers vary. Most boards want to hear that you are ready to take on the duties and responsibilities of a soldier in the next higher rank, that you have a general idea of what those specific responsibilities are, and that you have the basic knowledge to take care of soldiers and get the missions done.

Some board presidents will ask you to tell the board a little about yourself. Begin with when and where you were born and where you went to school. Then talk about when and where you entered the Army, your major assignments, your present job, your marital status, what you are doing to improve yourself, and what your goals are. Be honest. Tell the truth, not what you think the board wants to hear. If asked an opinion question, make sure you answer with your own opinion.

Before some boards you may be asked to name your goals, as well as your plans for a career or what you are currently doing to enhance your chances for promotion. If you actively work in these areas in advance, your answers will ring true. If you do not, you will probably be outdone by someone who has.

Prepare yourself for current events questions. Watch the local and national news on television for several days before appearing before the board. And be sure to read the local newspaper the day before and the morning of the board. Read your local post's newspaper if there is one.

You should have some knowledge of your unit's history. If you don't know the background of your unit, see your first sergeant.

When you report to the board, knock on the door to the boardroom and enter when instructed to do so. After entering, walk in the most direct route to a position centered on and two paces away from the table where the board president is seated. When appropriate, use facing movements, as in marching. The better your movements are executed, the more points you will receive for military bearing, so practice reporting to the board in advance.

Once you have stopped and are at the position of attention, render the hand salute and report to the president of the board saying, "Sergeant Major ('Sir,' if it is an officer), (your rank and name) reports to the president of the board as directed." Hold your salute until the board president returns it. Execute your salute in a snappy manner with very distinct moves. Some sergeants major will not have you salute when you report. Most will because they want to see you execute the movement.

After that, follow the directions of the board president. Some will have you make facing movements; others will have you sit down right away.

When you are told to be seated, you have two options. Both are acceptable. The first is to do an about-face, step to the chair, do another about-face, and then take your seat. The second option is to look to your rear, locate the chair intended for you, step backward to the chair, and take your seat. The first option is likely to be considered more military and impressive.

Once seated, sit erect with your feet on the floor and your hands resting in your lap or on your thighs. Most likely, the board president will tell you to relax. Try to do so, but maintain an erect military posture.

When answering a board member's questions, prefix all your answers with the board member's appropriate rank (sergeant, first sergeant, or sergeant major). Use "Sir" if it is an officer. Answer questions as accurately, honestly, and concisely as you can. Do not jeopardize yourself by giving answers you do not know to be correct. An incorrect answer may be worse than an honest, "Sergeant, I don't know the answer to that."

Direct your answer to the person asking the question, maintaining eye contact at all times. Often it is helpful to repeat the question as part of your answer (example: "Sergeant, the four lifesaving steps are . . ."). Speak loudly; it gives the air of confidence and can help to hide your nervousness.

When you are dismissed, stand, come to the position of attention, exchange salutes with the board president if the president is an officer, and step off, as in marching, on the most direct route for the door. Once outside the room, leave the building. Do not stop to exchange questions and answers with other soldiers waiting to go in.

Normally, you cannot be notified of the board's results until all board proceedings are over and the appropriate commander has signed all paperwork. The board only recommends; the appropriate commander authorizes your name to go on the promotion list.

Remember that the board will vote on your overall board performance: appearance, actions, oral expression, and types of questions you get right and wrong. How well you do will depend a great deal on how well you prepare yourself. Because your chain of command shares some responsibility for your preparation, they should assist you. But it is you going before the board and your promotion that is on the line.

Other Local Boards

Other local boards are also held: soldier or NCO of the month is one type, and competition for a school seat is another. Preparation for these boards is the same as for a promotion board. One great difference between the two is that, by regulation, promotion boards are limited to questions only. Other boards can have "hands-on" requirements, such as common tasks or a PT test, and they can be broken down into several phases that may take more than one day to complete.

DA Selection Boards

Boards are convened at the DA level to consider senior NCOs for promotion, schools, separation, and assignments. At these boards, your records are what count. The image that your photo and written history projects to the board members will make or break you. A member of a DA board does not have the time or ability to second-guess your records. Because of this, soldiers who perform poorly have been promoted because their records indicated otherwise. Likewise, good soldiers have not been promoted because their records were not complete.

Don't assume that a clerk in your local personnel office will take care of you. You control what the board will see by your daily and yearly conduct, by how closely you screen your records, and by your physical appearance as represented on your official photograph. It is critical that you are careful when reviewing your records and when having a DA photo taken. DA selection boards continue to receive record files with the following problems:

- No photograph or one that is not current
- Photographs showing uniforms that don't fit, have missing awards and decorations, or ones that are worn incorrectly.
- Information in files that is not current or just plain inaccurate.

You may be the sharpest soldier in your unit, but if your records aren't in order, you are going to have problems.

There is no excuse these days for not being ready for a selection board. Today's soldiers have printed books and manuals and a virtual library online. Through Web sites such as Army Knowledge Online *(www.army.mil)*, soldiers can stay up-to-date on any subject involving the military.

Board Quick Tips

1. Start studying for a board the day you find out you are going to appear before one.
2. Don't study alone. Studying with another soldier will help you both to increase your knowledge.
3. Practice drill movements (about-face, left face, right face, and stepping off).
4. Practice answering in a clear, firm voice.
5. Have your uniform cleaned and pressed.
6. Have your uniform checked by your platoon sergeant and first sergeant two to three days before the board (no later than one day before the board).
7. Don't cram the night before the board. If you don't know the basic subject matter by then, cramming will not help.
8. Read the paper and watch TV news programs the day before and the morning of the board.
9. Rest well the night before the board.
10. Eat a light breakfast the morning of the board.
11. After the board, critique yourself. Write down or highlight in your study guide as many of the questions as you can remember. Find the answers to the questions you missed or had trouble with.

One last caution: don't disregard any training or information as too old or not necessary. I have heard young soldiers say things like, "I'm not in combat arms and don't need to know all that stuff," or "We have GPS systems now and that makes learning land navigation obsolete." Yes, the Army does have a lot of new equipment that makes many things easier, and every soldier is not in combat or combat support units. However, commanders study tactics of battles fought hundreds and hundreds of years ago for a reason. The more you know the better you can adapt when the need comes. You had better know how to find your way if your GPS system goes out.

4

Sample Questions

Following are sample questions that I have collected over the past years. They cover most of the subject areas that are common to boards. Although most will be found if you search through the manual or manuals covering the subject, some questions will not be found in current manuals. These examples are based on general military information or old material from manuals that refuse to die. They have been asked on boards for years, and their information is still valid.

Don't be surprised if you hear a question not covered in this book. It would be almost impossible to cover every subject for every military specialty in the Army, so we only cover the basics. The bottom line is that if you can answer the questions from this guide, you will have a solid background of military knowledge from which to lead, work, and get ahead by doing well on boards.

At the end of many of the subject areas, there is space for you to add questions and answers as you come across them. By updating your guide with new questions, you will be prepared to appear before and to serve on local boards at a later date.

Remember that a question on any given subject can be asked in several different ways. *Learn, don't memorize!* You should be told the subject area from which the question is coming. Listen for keywords that will give you an idea of what the questioner is looking for. For example, a keyword could be "fracture." You could be asked, "What two types of fractures are there?" The board member asking the question could be looking for "open and closed" or "simple and compound." Either answer could be accepted as correct. If you know about fractures, you'll have the answer no matter how the question is asked. Take your time. Think about the questions, then think about your answers.

Army Command Policy
REFERENCE: AR 600-20, Army Command Policy

1. According to AR 600-20, the NCO support chain assists the chain of command in ten areas. Name five of them.
   (1) Transmitting and instilling the professional Army ethic.
   (2) Planning and conducting day-to-day unit operations.
   (3) Training of enlisted soldiers in MOS and basic skills.
   (4) Supervising physical fitness training.
   (5) Teaching military customs, courtesies, traditions, and history.
   (6) Caring for individual soldiers.
   (7) Teaching the mission of the unit and training to accomplish it.
   (8) Accounting for and maintaining individual arms and equipment.
   (9) The NCO development program.
   (10) Achieving and maintaining courage, candor, competence, and commitment.

2. What AR covers the relationship between soldiers of different rank?
   AR 600-20.

3. AR 600-20 defines sexual harassment as what?
   Gender discrimination that involves unwelcomed sexual advances, requests for sexual favors, and other conduct of a sexual nature.

4. What behaviors constitute sexual harassment?
   Verbal comments.
   Nonverbal gestures.
   Physical contact.

5. What is the purpose of the Army Sexual Assault Prevention and Response Program?
   • *To eliminate incidents of sexual assault.*
   • *To hold perpetrators accountable for their actions.*
   • *To treat victims with dignity, fairness, and respect.*

6. What tools are used in the Army Sexual Assault Prevention and Response Program?
   - *Education and training.*
   - *Awareness and prevention.*
   - *Victim advocacy.*
   - *Incident reporting.*
   - *Accountability for actions.*

7. Where and when does the Army Sexual Assault Prevention and Response Program apply?
   - *On and off post.*
   - *During duty and non-duty hours.*
   - *To all working, living, and recreational environments.*

8. What acts fall under the label of sexual assault?
   - *Unwanted or inappropriate sexual conduct or fondling.*
   - *Rape.*
   - *Non-consensual oral or anal sex.*

9. Who is responsible for the prevention of sexual harassment?
   The commander.

10. Name some of the programs or policies covered by AR 600-20.
    - *Equal Opportunity and Prevention of Sexual Harassment Programs.*
    - *The Sexual Assault Prevention and Response Program.*
    - *The Army Well-Being process.*
    - *Relationships between soldiers of different ranks.*
    - *Guidance on government travel cards.*
    - *Domestic Violence Amendment to the Gun Control Act of 1968.*
    - *Installation Command and Control.*
    - *The NCO Support Channel.*
    - *Hazing.*

11. What is Army Well-Being?
    It is the physical, material, and spiritual states of the Army family.

12. What is the Army Family?
    The Army Family is made up of the soldiers (active, reserve, and guard), retirees, veterans, DA civilians, and their families that contribute to preparedness and support the Army Mission.

13. What is the Army's policy on hazing?
    Hazing is in opposition to Army values and is prohibited.

14. What is hazing?
    Hazing is conduct whereby one military member or employee causes another military member or employee to suffer or be exposed to an activity that is cruel, abusive, oppressive, or harmful.

15. What is the commander required to do to a soldier with AIDS?
    Order the soldier that is HIV-positive to inform any sexual partner of the infection prior to engaging in intimate sexual behavior.

16. What is the Army policy on sexual harassment?
    That sexual harassment is unacceptable conduct that will not be tolerated.

17. How do commanders and leaders earn the loyalty of their soldiers?
    By showing loyalty to soldiers, the Army, and the nation.

18. How do commanders and leaders build a positive command climate?
    By considering soldiers' needs, caring for their well-being, and demonstrating genuine concern.

Army Programs
REFERENCE: Various

1. Name ten Army programs.
   Army Community Services (AR 608-1).
   Army Continuing Education System (AR 621-5).
   Army Emergency Relief (AR 930-4).
   Army Equal Opportunity Program (AR 600-200).
   Army Family Team Building Program (AR 608-48).
   Army Lessons Learned Program (AR 11-33).
   American Red Cross (AR 930-5).
   Army Reenlistment/Retention Program (AR 601-280).
   Army Safety Program (AR 385-10).
   Army Sponsorship Program (AR 600-8-8).
   Army Substance Abuse Program (AR 600-85).
   Army Quality of Life Program.
   Army Weight-Control Program (AR 600-9).
   Better Opportunities for Single Soldiers (DA Cir 608-04-1).
   Noncommissioned Officer Development Program
      (DA Pam 600-25).
   Army Sexual Assault Prevention and Response Program
      (AR 600-20).

2. What is the Army's Quality of Life Program?
   A group of policies, programs, and actions designed to develop military commitment and cohesiveness essential to the combat readiness of the Army.

3. What are the six subsystems of the Quality of Life Program?
   (1) Army Community Service (ACS).
   (2) Morale and Support Activities.
   (3) Army Continuing Education System (ACES).
   (4) Army Club Management System.
   (5) Army Postal Services.
   (6) Army and Air Force Exchange Services.

4. The Quality of Life Program has programs and assistance categorized under two general headings—what are they?
   Living conditions and duty environment.

5. What does AFAP stand for and what is it?
Army Family Action Plan; a program to support healthy family functioning by preventing child and spouse abuse through identification, treatment, and rehabilitation.

6. What does the Army Family Action Plan do?
Alerts commanders and leaders to areas that need their concern.
Safeguards well-being.
Helps retain soldiers.
Enhances the Army image.

7. What are the objectives of the Army Morale, Welfare, and Recreational program?
   • *Support combat readiness and effectiveness.*
   • *Support reenlistment and retention of quality personnel.*
   • *Provide leisure time activities to support a quality of life commensurate with American values.*
   • *Promote mental and physical well-being of personnel.*
   • *Foster community pride, soldiers' morale, family wellness, and unit esprit de corps.*
   • *Erase the impact of such things as relocations and deployments.*

8. What does the Better Opportunities for Single Soldiers (BOSS) provide?
   • *Physical activities.*
   • *Self-development activities.*
   • *Leisure activities.*
   • *Educational activities.*

9. The BOSS program places emphasis on what group of soldiers?
Single and unaccompanied.

10. What is the purpose of the Lessons Learned program?
The purpose is to sustain, enhance, and increase the Army's preparedness to conduct current and future operations through soldiers' collecting positive and negative information and submitting it through their chain of command.

11. What is the goal of the Sexual Assault Prevention and Response program?
The goal is to eliminate incidents of sexual assault through aware-ness and prevention, training and education, victim advocacy, response, reporting, and accountability.

REFERENCE: AR 600-85, Army Substance Abuse Program (ASAP)

1. In a nutshell, what are the main objectives of the ASAP?
   • *Prevent alcohol and drug abuse.*
   • *Identify abusers as early as possible.*
   • *Rehabilitate and restore abusers to effective duty.*
   • *Separate rehabilitation failures from the service.*

2. ASAP has a number of objectives. Name them.
   (1) Prevention and deterrence.
   (2) Identification.
   (3) Rehabilitation.
   (4) Education.
   (5) Treatment.
   (6) Improve unit readiness.
   (7) Increase individual fitness.

3. Who is authorized to use ASAP?
   • *Active-duty soldiers and their family members.*
   • *Retired soldiers and family members.*
   • *Reserve soldiers and their family members.*
   • *DOD civilians and their family members.*
   • *Army National Guard and Reserve personnel on a space/resource available basis.*

4. Within how many days after arrival in a new duty station should a soldier be educated on ASAP?
   Sixty days.

5. List the five means of identification or referral to ASAP.
   (1) Voluntary (self-referral).
   (2) Command referral.
   (3) Medical referral.
   (4) Apprehension.
   (5) Urinalysis program (biochemical).

6. State some items you should include when counseling soldiers on ASAP.
   - *Availability.*
   - *Referral procedures.*
   - *Location.*
   - *Types of treatment available.*
   - *Punishment under Uniform Code of Military Justice (UCMJ).*
   - *Separation procedures.*

7. Can a soldier enroll in ASAP to avoid pending disciplinary action?
   No.

8. Can a soldier enrolled in ASAP reenlist?
   No, but a soldier can extend his or her service long enough to complete the program and then apply to reenlist.

9. Mention several indicators of a drug or alcohol problem.
   - *Any major consistent change in a soldier's normal perfor-mance, such as failure to show for duty or tardiness.*
   - *Signs of a hangover.*
   - *Fighting.*
   - *Deterioration in personal appearance.*
   - *Extended lunch hours.*
   - *Reduced work performance.*
   - *Intoxication at off-duty functions.*
   - *Blackouts.*
   - *Abnormal behavior.*
   - *Odor of alcohol in the morning.*

10. What is the Army's policy on alcohol?
    To maintain a workplace free from alcohol.

11. What are the three main categories of drugs?
    Hallucinogens, depressants, and stimulants.

12. What does the ASAP program emphasize?
    Readiness and personal responsibility.

13. If you become aware of drug use in the unit, what is your responsibility?
    To report any drug use to the chain of command.

14. At what point during the rehab process is an alcoholic considered cured?
    Never, alcoholism is an incurable disease. The alcoholic can only be considered as recovering.

15. Who makes the final decision as to a rehabilitative success or failure?
    Both the unit commander and the ASAP counselor.

16. When is a soldier considered a rehab failure?
    When he or she is unable or unwilling to be returned to effective duty after short-term rehab efforts.

17. Can the fact that a soldier had been enrolled in ASAP be revealed without the soldier's permission?
    No.

18. Is a Line of Duty determination required during detoxification?
    No.

19. What is the Unit Prevention Leader's role in the ASAP?
    - *Design, implement, and evaluate the unit's ASAP plan.*
    - *Assist in briefing all new personnel on ASAP.*
    - *Administer the unit biochemical testing program.*
    - *Keep the commander informed on ASAP trends.*
    - *Develop, coordinate, and deliver prevention education material and training.*
    - *Be a liaison with the ASAP counseling center or medical unit when deployed.*

20. What rank must the UPL be?
    Officer or NCO that is an E5 or above.

21. What are the rehabilitation objectives of the ASAP?
    - *Return soldiers to full duty as soon as possible.*
    - *Identify soldiers who cannot be rehabilitated.*
    - *Assist and refer soldiers who are discharged to a treatment facility in the vicinity of where they reside.*
    - *Help resolve substance abuse problems in the families of soldiers.*

22. What are some of the ASAP principles?
    - *Substance abuse is inconsistent with Army values and the standards of performance, discipline, and readiness.*
    - *Unit commanders must intervene early.*
    - *ASAP is mandatory for all command-referred soldiers.*
    - *Substance abusers may be enrolled in ASAP when clinically recommended.*
    - *Soldiers who fail to participate or respond successfully to the program will be processed for administrative separation.*
    - *Alcohol and drug abuse will be addressed in a single program.*
    - *Unit commanders retain their authority to make personnel decisions.*
    - *The confidential nature of ASAP treatment records will be preserved.*
    - *Prevention and education will be given the highest priority.*

23. What are some of the prevention policies in the ASAP?
    - *Alcohol deglamorization.*
    - *Education and training on the consequences of substance abuse.*
    - *Early identification of substance abusers.*
    - *Prevention initiatives within the total community.*
    - *Substance abuse training conducted throughout the Army Training System.*

24. What is the minimum amount of annual substance abuse and awareness training a soldier is to receive?
    Four hours.

25. Who may authorize alcohol consumption during duty hours?
    The first General Officer/installation commander in the chain of command.

26. What form is signed by the Commander or 1SG to make an ASAP referral?
    DA Form 8003—Referral Form.

27. What do the letters ASAP stand for?
    Army Substance Abuse Program.

REFERENCE: AR 930-5, American Red Cross (ARC)

1. Describe several services that the American Red Cross provides for soldiers and their families.
   - *Financial assistance in the form of a loan without interest on the basis of need.*
   - *Referral services to agencies specializing in legal aid, medical or psychiatric care, employment, or family and children's welfare.*
   - *Instructional courses such as water safety, first aid, home care of the sick or injured, and mother and baby care.*
   - *Communication between servicemembers and their families to alleviate apprehension, worry, and concern; information for justification of emergency leave, deferments, compassionate reassignments, and leave extension requests.*
   - *Volunteer assistance programs.*
   - *Blood programs.*

2. What is the major service the American Red Cross provides to servicemembers?
   Communication between servicemembers and their families for problem solving and emergencies.

3. How is the Red Cross primarily staffed?
   By volunteers.

4. What services does the Red Cross provide for prisoners of war (POWs)?
   It handles inquiries concerning welfare, obtains names of POWs, and aids in the packaging and distribution of health and welfare packages.

5. Why do the Army Emergency Relief (AER) and the American Red Cross coordinate their efforts?
   To avoid duplication of effort and to assure coverage of all areas of aid.

6. How long has the Red Cross been in existence?
   Since 1905.

REFERENCE: AR 608-1, Army Community Services (ACS)

1. What services does the ACS provide for the soldier?
   - *Army Family Advocacy Program.*
   - *Foster Care Program.*
   - *Exceptional Family Members Program.*
   - *Assistance in relocating families.*
   - *Loan of household items.*
   - *Budget, debt, and financial counseling.*
   - *Employment assistance for spouses.*
   - *Recreational activities.*
   - *Child care facilities.*
   - *Babysitting services.*
   - *Welfare resources.*

2. When was the ACS established?
   1965.

3. What is the ACS theme?
   Self-help, service, and stability.

REFERENCE: AR 621-5, Army Continuing Education System (ACES)

1. What are the two goals of ACES?
   To offer soldiers the opportunity for continuing education to achieve Army and personal education goals, and to help family members achieve their personal educational goals whenever possible.

2. List several services offered under ACES.
   - *Counseling.*
   - *Testing.*
   - *Basic Skills Education Program (BSEP).*
   - *Advanced Skills Education Program (ASEP).*
   - *MOS-related classes.*
   - *Resource center.*
   - *General Educational Development (GED) completion.*
   - *General Technical (GT) improvement program.*
   - *College prep classes.*

- *Associate, baccalaureate, and graduate degrees.*
- *Defense Activity for Non-Traditional Education Support (DANTES).*

3. What are the four common programs under ACES?
   (1) Academic education.
   (2) Skill development.
   (3) Skill recognition.
   (4) Education services.

4. What is the CLEP program?
   College Level Examination Program, which enables students to earn college credit by successful completion of a course examination.

5. What is DANTES?
   Defense Activity for Non-Traditional Education Support. It is a distance learning program that allows soldiers to take college or technical courses from accredited schools by correspondence, computer conferencing, and classroom.

REFERENCE: AR 930-4, Army Emergency Relief (AER)

1. What is the purpose of AER?
   To help Army members and their families in emergency financial situations by "helping the Army take care of its own" (motto).

2. List three types of financial assistance that the AER provides.
   Loans without interest, grants, and a combination of loans and grants.

3. How does the AER fund its programs?
   Through voluntary contributions from Army personnel during the AER fund campaign, through unsolicited contributions, and through income from investments.

4. What form is used to apply for an AER loan?
   DA Form 1103.

5. If a soldier needs an AER loan, who initiates the paperwork?
   The unit/command financial adviser.

6. Name several situations where the AER is likely to give a loan.
   - *Medical and dental expenses.*
   - *Loss of funds.*
   - *Nonreceipt of pay.*
   - *Repair of privately owned vehicle (POV) if essential transportation.*
   - *Rent deposits.*
   - *Funeral expenses.*
   - *Emergency travel expenses.*
   - *Cost of fire or other disaster.*
   - *Pay for food and utilities.*

7. Name some situations where the AER will not give a loan.
   - *Marriage funds.*
   - *Ordinary leave.*
   - *Court fees.*
   - *Convenience or luxury items.*
   - *To pay for overdrawn bank accounts.*
   - *Divorces.*
   - *Liquidation of debts.*

8. Are contributions to the AER tax deductible?
   Yes.

Tricare

1. What is Tricare?
   Tricare is a regionally managed healthcare program for active-duty and retired soldiers and their families or survivors.

2. Name the three Tricare options for active-duty soldiers.
   Tricare Prime, Tricare Extra, and Tricare Standard.

3. What is the enrollment fee for active-duty members in Tricare?
   There is no fee for active-duty members and their families.

4. Specify the requirement for active-duty members to receive Tricare.
   Completion of an enrollment form.

REFERENCE: AR 600-20, Army Command Policy (Equal Opportunity)

1. Equal opportunity in the Army is based on what?
   Fairness, justice, and equity.

2. Fair treatment in the Army is based on what?
   Merit, fitness, capability, and in support of readiness.

3. What is the purpose of the EO program?
   To provide equal opportunity for military personnel and family members; and to create and sustain effective units by eliminating discriminatory behaviors or practices.

4. What is the definition of personal racism or sexism?
   The acting out of prejudices by individuals against other individuals or groups based on their race or gender.

5. What is institutional discrimination?
   The policies, practices, and procedures of an institution that result in unequal treatment because of racial, religious, gender, economic, or social differences.

6. How many periods of EO training are soldiers required to receive each year?
   Four. Two on EO topics and two on prevention of sexual harassment.

7. What is the primary source for solving EO complaints?
   The chain of command.

8. Who is responsible for the EO program in your unit?
   The unit commander.

9. Unit Equal Opportunity Representatives (EORs) are required to be of what rank?
   E5 (P) through 1LT.

10. All EO complaints should be what?
    Filed in person when possible.

11. What are individuals responsible for in the EO program?
    - *Reporting any incident of sexual harassment and discrimination and providing the command with an opportunity to take appropriate action.*
    - *Submitting only legitimate complaints.*

12. What action does the Army take against civilian establishments that discriminate against soldiers and their families?
    Place them off limits.

13. Are all units required to have an EO representative?
    Yes.

REFERENCE: AR 600-8-8, Army Sponsorship Program

1. Explain the purpose of the Army Sponsorship Program.
   To assist soldiers and their family members in getting established at a new duty station.

2. List the three types of sponsorship programs.
   (1) In-sponsorship.
   (2) Out-sponsorship.
   (3) Reactionary sponsorship.

3. What is the in-sponsorship program?
   The normal activities, information, and assistance provided to newly arrived soldiers and their family members.

4. What is the out-sponsorship program?
   The assistance provided to soldiers and their family members who are departing a unit.

5. When a gaining commander receives a request for sponsorship, a welcome packet must be sent within how many days?
   Ten days.

6. Name some things a sponsor should do prior to an individual's arrival.
   - *Send a welcome letter.*
   - *Furnish a map of the area.*

- *Provide information on community services and on-post facilities.*
- *Provide information on the surrounding area.*

7. How long must a soldier be on temporary duty (TD) before he or she can participate in the advanced arrival program?
Ninety days.

8. What grade must a sponsor be?
A grade equal to or higher than that of the individual he or she is sponsoring.

9. State the two degrees of sponsorship.
Full and limited.

10. Why is it important to have an active and well-managed sponsorship program?
To create a positive first impression for new soldiers and to decrease distractions that hamper personal performance and mission accomplishment.

11. What form is used to transmit sponsorship requirements?
DA Form 5434.

12. What actions should a sponsor take?
- *Send a welcome letter.*
- *Use ACS services.*
- *Greet the incoming soldier.*
- *Provide an orientation.*
- *Assist with inprocessing problems.*

REFERENCE: AR 600-9, The Army Weight Control Program

1. What are the overall objectives of the AWCP?
To see that soldiers of all ranks are able to meet the physical demands of their duties under combat conditions and that they present a trim military appearance at all times.

2. What are the general objectives of the AWCP?
To assist in establishing and maintaining discipline, operational readiness, optimal physical fitness, health, and effectiveness.

3. How often are soldiers to be weighed under the AWCP program?
   When they take a physical fitness test or at least every six months.

4. How does being placed in the AWCP affect a soldier's life?
   * *They are non-promotable*
   * *They cannot be assigned to Command, Command Sergeant Major, or First Sergeant positions.*
   * *They cannot attend professional military schools.*

REFERENCE: AR 600-8-101, Personnel Processing

1. What stations are required for all soldiers during in- and out-processing?
   Personnel Information Station (PINS)
   Personnel Strength Management Station
   Personnel Management Station
   Medical facility
   Tricare Service Center
   Dental facility
   Defense Enrollment Eligibility System (DEERS)
   Identification station (card and ID)
   Security office
   Education Center
   Military Pay

Awards and Decorations

REFERENCES: AR 670-1, Wear and Appearance of Army Uniforms
and Insignia; AR 600-8-22, Military Awards

1. What is an award?
   A decoration, medal, badge, ribbon, or appurtenance bestowed on an individual or unit.

2. Who can make recommendations for awards?
   Any person having knowledge of an outstanding act or achievement.

3. What is the time limit for making a recommendation for an award?
   Within two years of the act, achievement, or service.

4. Name the movie star who was the most highly decorated soldier in World War II.
   Audie Murphy.

5. Describe a decoration.
   An award to an individual for a specific act of gallantry or service.

6. Specify the two categories of decorations.
   Heroism and achievement.

7. Describe fourragére.
   A rope-like device worn to signify the award of a foreign decoration.

8. May an enlisted person be awarded the Legion of Merit?
   Yes.

9. What medal, still being awarded today, was the first medal awarded in the U.S. Army? For what was it initially awarded?
   The Purple Heart. Initially, it was awarded for valor in the Revolutionary War.

10. To whom is the Purple Heart awarded today?
    To anyone, military or civilian, wounded or killed in action against an armed enemy.

11. What decoration has George Washington's picture on it?
    The Purple Heart.

12. Name the three highest awards.
    Medal of Honor, Distinguished Service Cross, and the Defense Distinguished Service Medal.

13. What is the highest award given by the Army in time of war?
    The Medal of Honor.

14. What is meant by "above and beyond the call of duty?"
    The display of exemplary courage or professionalism in the acceptance of danger or extraordinary circumstances, which is not normally expected of a person.

15. Who is the approving authority for the Medal of Honor?
    The U.S. Congress.

16. Who must award the Medal of Honor?
    The President of the United States.

17. Before the Medal of Honor was established in 1862, what award was given for personal bravery or self-sacrifice?
    The Purple Heart.

18. What is the second highest award for valor?
    The Distinguished Service Cross.

19. What is the highest peacetime award?
    The Distinguished Service Medal.

20. What is the only peacetime decoration awarded for valor?
    The Soldier's Medal.

21. For what is the Soldier's Medal awarded?
    For distinguishing oneself by heroism not involving conflict with an armed enemy.

22. What is engraved on the back of the Medal of Honor?
    The grade, name, and unit of the person receiving the award.

23. What award is given for exemplary behavior, efficiency, and fidelity during enlisted status and active duty?
    The Good Conduct Medal.

24. Who is the approving authority for the Good Conduct Medal?
    The company commander.

25. What words are engraved on the Good Conduct Medal?
    Honor, efficiency, and fidelity.

26. Who is the approving authority for the Army Achievement Medal?
    The battalion commander.

27. How often can you receive a good conduct award?
    Every three years.

28. What are the four categories of badges worn on the Army uniform?
    (1) Marksmanship.
    (2) Combat and special skill.
    (3) Identification.
    (4) Foreign.

29. What is the maximum number of foreign badges that can be worn on the uniform at any time?
    One.

30. What is the maximum number of marksmanship badges that can be worn on the uniform at any time?
    Three.

31. What is the meaning of the Combat Infantryman's Badge (CIB) with the star?
    Two awards of the CIB.

32. Describe appurtenances.
    Devices affixed to service ribbons and suspension ribbons for decorations, medals, and awards. They are worn to denote an additional award, participation in a specific event, or some other distinguishing characteristic of an award. Examples are oak leaf clusters, "V" devices, numbers, clasps, or stars.

33. What is the requirement for awarding the Army Service Ribbon?
    Successful completion of an initial MOS-producing course.

34. When will an individual be authorized to first wear the NCO Professional Development Ribbon?
Upon completion of the Primary Leadership Development Course (PLDC).

35. What is the main difference between wearing a unit award and wearing an individual award?
A unit decoration is worn only when you are serving with that unit, unless you were present (assigned or attached on orders) when the unit won the award.

36. What are the awards given to Army units?
(1) Presidential Unit Citation.
(2) Valorous Unit Award.
(3) Joint Meritorious Unit Award.

37. What is the time limit for recommending an award?
Two years.

38. What are the four special skill tabs?
(1) Special Forces Tab.
(2) Ranger Tab.
(3) Sapper Tab.
(4) Presidents Hundred Tab.

39. What are the three U.S. Army Combat Badges?
(1) CIB—Combat Infantryman's Badge.
(2) CAB—Combat Action Badge.
(3) CMB—Combat Medical Badge.

40. May a soldier be awarded more than one combat badge?
Yes, but not for the same qualifying period.

41. How many "V" devices can be worn on a ribbon?
One.

42. What is the goal of the Army Awards Program?
To foster mission accomplishment through recognizing excellence and motivation for performance and service.

43. What is the objective of the Army Awards Program?
    To provide tangible recognition for acts of valor, exceptional service or achievement, special skills or qualifications, and acts of heroism not involving actual combat.

44. Who was the first Medal of Honor recipient?
    Private Jacob Parrott, Company K, 33d Ohio Volunteers

Basic Combat Skills

REFERENCES: FM 3-21.75, The Warrior Ethos and Soldier Combat Skills;
TRADOC PAM 600-4, IET Soldier's Handbook; FM 7-22.7, The NCO Guide

1. Who is responsible for military intelligence?
   Every individual.

2. Name the three Ws of an observation report.
   What, when, and where.

3. What does the acronym SALUTE stand for?
   —Size.
   —Activity.
   —Location.
   —Unit.
   —Time.
   —Equipment.

4. What are some common signatures associated with a camouflaged command post?
   • *Concentration of vehicles.*
   • *Antennas.*
   • *Converging common lines.*
   • *Protective wire barriers.*
   • *New access routes or trails.*

5. What is the mission of a quartering party?
   To reconnoiter, to aid in preparation and occupation of positions, and to ease supply problems.

6. What is the purpose of a stand-to?
   To ensure that every soldier adjusts to the changing light and noise conditions, and to ensure that all are dressed, equipped, and ready for action.

7. Define the keyword OCOKA.
   O—Observation and field of fire.
   C—Cover and concealment.
   O—Obstacles.
   K—Key terrain.
   A—Avenues of approach.

8. What are the procedures for sending messages?
   - *Routine.*
   - *Priority.*
   - *Immediate.*
   - *Flash.*
   - *Emergency Command Precedence.*

9. Name the five means of communication.
   (1) Radio.
   (2) Visual signals.
   (3) Sound.
   (4) Wire.
   (5) Messenger.

10. What information is found in a signed operation instruction (SOI)?
    Current call signs, suffixes, and frequencies.

11. What are the seven steps in the recommended approach for training?
    (1) Set the objectives.
    (2) Plan the resources.
    (3) Train the trainers.
    (4) Provide the resources.
    (5) Assess risk and safety concerns.
    (6) Conduct the training.
    (7) Evaluate the results.

12. Name the paragraphs in an operation order.
    - *Situation.*
    - *Mission.*
    - *Execution.*
    - *Service and support.*
    - *Command and signal.*

13. What are the key steps in the troop-leading procedure?
    - *Receive the mission.*
    - *Issue a warning order.*
    - *Make a tentative plan.*
    - *Start necessary movement.*
    - *Reconnoiter.*

- *Complete the plan.*
- *Issue the complete order.*
- *Supervise.*

14. What is a maneuver?
    Movement, supported by fire, to a position of advantage from which you can destroy or threaten the enemy.

15. What is a patrol?
    A unit sent out to perform an assigned mission of reconnaissance, combat, or a combination of the two.

16. What are the four key principles of successful patrolling?
    (1) Detailed planning.
    (2) Thorough reconnaissance.
    (3) Position control.
    (4) All-round security.

17. Mention some types of combat patrols.
    - *Raid.*
    - *Reconnaissance.*
    - *Ambush.*
    - *Contact.*
    - *Search and destroy.*
    - *Economy of force.*

18. What are the elements of a combat patrol?
    - *Assault.*
    - *Security.*
    - *Support.*
    - *Headquarters.*

19. What is the purpose of a recon patrol?
    To provide the commander with timely, accurate information on the enemy and the terrain he controls.

20. Name the two kinds of recon patrols.
    Point and area.

21. What are the elements of a recon patrol?
    Reconnaissance, security, and headquarters.

22. What are the three types of rallying points?
   Initial, en route, and objective.

23. Why should a soldier never carry unnecessary equipment?
   Extra weight slows down progress.

24. What is a retrograde operation?
   Organized movement to the rear, away from the enemy.

25. List the three types of retrograde action.
   Delay, withdrawal, and retirement.

26. What is combat power?
   A unit's ability to fight.

27. List the five basic rules of combat.
   (1) Secure.
   (2) Move.
   (3) Shoot.
   (4) Communicate.
   (5) Sustain.

28. Where does the strength of a unit come from?
   The skill, courage, and discipline of the individual soldiers.

29. What are the four Fs of fighting?
   (1) Find them.
   (2) Fix them.
   (3) Fight them.
   (4) Finish them.

30. What are the two types of defense?
   Area and mobile.

31. Identify the task first in priority when organizing for the defense.
   Establishing security.

32. What are field fortifications?
   Temporary shelters providing protection from enemy fire.

33. List some examples of artificial cover.
    Foxholes, trenches, and walls.

34. What three things should be considered when selecting a temporary fighting position?
    Observation, field of fire, and cover and concealment.

35. Why is it important to furnish your fighting position with overhead cover?
    To protect against mortar rounds and fragments from other indirect fire weapons.

36. What is a field of fire?
    An assigned area that a soldier is to cover with fire from his or her position.

37. What information is marked on a standard range card?
    • *Sketch section.*
    • *Data section.*
    • *Sectors of fire.*
    • *Military symbols.*

38. What must a range card depict?
    • *Sectors of fire.*
    • *Prominent terrain features and dead space.*
    • *Weapons symbols.*
    • *Targets and their ranges.*
    • *Final protective line (FPL).*
    • *Principal direction of fire (PDF).*
    • *Marginal data.*

39. Name the six standard positions for firing an M16 rifle.
    (1) Foxhole.
    (2) Prone.
    (3) Prone, supported.
    (4) Kneeling.
    (5) Kneeling, supported.
    (6) Standing.

40. What is used to protect soldiers (riflemen) from frontal, small-caliber direct fire and limited fragmentation?
Individual hasty and deliberate fighting positions with overhead cover.

41. When you are in a defensive perimeter, why is it recommended that you use grenades whenever possible?
Their bursting radius makes them effective against uncertain targets, and they do not disclose your position.

42. What are the two ways to detonate an explosive charge?
Electric and nonelectric.

43. What is the difference between military and commercial dynamite?
Military dynamite is made with RDX (cyclonite) and is not shock-sensitive like dynamite made with nitroglycerin.

44. How many inches of timber does it take to stop small-caliber rounds, up to those from a 7.62mm machine gun?
36 inches.

45. How many inches of reinforced concrete does it take to stop small-caliber rounds, up to those from a 7.62mm machine gun?
Six inches.

46. A roll of concertina wire should never be extended wider than how many meters?
15 meters.

47. What are the four main characteristics of a nuclear explosion?
(1) Blast.
(2) Thermal radiation.
(3) Nuclear radiation.
(4) Electrical magnetic pulse (EMP).

48. List the six classes of fire with respect to a machine gun.
(1) Fixed.
(2) Traverse.
(3) Searching.
(4) Traverse and searching.
(5) Swing traverse.
(6) Free gun.

49. When firing a machine gun, what is meant by searching fire?
Moving the muzzle of the weapon up and down to deliver fire in depth across a target.

50. What is the primary reason for placing a tracer after every fifth round in a belt of machine-gun ammo?
Observation of fire and to quickly place rounds on the target.

51. What are active security measures?
Observation posts (OPs), listening posts (LPs), and guards.

52. What are passive security measures?
   * *Camouflage.*
   * *Controlling movement.*
   * *Light and noise discipline.*
   * *Limiting radio traffic.*

53. Describe a defilade position.
A position behind terrain, usually on the reverse slope of a hill, which completely masks the position from enemy observation.

54. Explain the three basic rules for camouflage.
Take advantage of all natural concealment; alter the form, shadow, texture, and color of objects; camouflage against air and ground observation.

55. In a tactical area, unless required to wear the helmet, why is it generally preferable to wear a soft cap?
The helmet has a distinct shape (silhouette), and it may muffle sounds, especially if there is a breeze.

56. What is evasion?
Action taken when isolated behind enemy lines to stay out of enemy hands and to get back to your own friendly lines.

57. Why are the high and low crawls not suitable when very near the enemy?
Crawling makes a shuffling noise that is easily heard.

58. About how long does it take for a soldier's eyes to become accustomed to the dark and allow him to distinguish objects in dim light?
About thirty minutes.

59. What does BMNT stand for?
Beginning (of) Morning Nautical Twilight.

60. What does EENT stand for?
End (of) Evening Nautical Twilight.

61. How should you react to a flare at night?
   • *Close one eye.*
   • *Assume a prone position.*
   • *Move out of the illuminated area.*
   • *Continue the mission.*

62. If you are separated from your unit during combat, what becomes your mission?
To rejoin your unit.

63. What does the acronym SERE stand for?
      —*Survival.*
      —*Evasion.*
      —*Resistance.*
      —*Escape.*

64. What are the two methods of holding a lensatic compass when sighting?
Compass to cheek and center-hold methods.

65. Name two methods of determining direction without a compass during daylight hours.
Shadow tip method and watch method.

66. What is escape?
Action taken to get away from an enemy force if you are captured.

67. When is the best possible time to escape if captured?
As soon as possible.

68. Describe the major advantages to early escape if captured.
    - *Closer to your own lines.*
    - *Fewer guards.*
    - *Inexperienced guards.*
    - *More confusion.*
    - *Better physical condition.*
    - *Better psychological aspects.*

69. What document governs your actions in case of capture?
    The Code of Conduct.

70. Can a prisoner of war be deprived of his rank?
    No.

71. Your fighting position must provide two things. What are they?
    Allow you to effectively fire your weapon and provide protection from direct and indirect enemy fire.

72. What is the height of grazing fire?
    One meter above the ground.

73. What are the first two things you do when building a machine-gun position?
    Mark the position of the tripod legs and then the sectors of fire.

74. What are the components of an Eagle cocktail?
    A plastic or rubberized bag; a gas and oil mixture; a smoke grenade; a thermite grenade; and string, tapes, or wire to tie up the bundle.

75. What are the weak points on an armored vehicle?
    Turret ring, suspension system, fuel tanks, engine compartment, and ammunition storage areas.

76. What does LCE stand for?
    Load carrying equipment.

77. In urban combat, how would you move parallel to a building?
    Use smoke for concealment, stay close to the side of the building, use shadows if possible, move quickly from covered position to covered position, and have someone overwatch my movement.

78. What are considered natural kill zones in an urban area?
    Streets, alleys, and parks.

79. What are the basic rules for entry into a building?
    • *Select an entry point before moving.*
    • *Avoid windows and doors.*
    • *Use smoke for concealment.*
    • *Make new entry points with explosives.*
    • *Throw a grenade through the entry point before entering.*
    • *Quickly follow the explosion of the grenade.*
    • *Have a buddy overwatch you as you enter.*
    • *Enter at the highest point possible.*

80. What is the best way to make a low-level entry into a building?
    • *Make a hole using explosives or tank round.*
    • *Cook off a grenade and toss it through the entry point.*
    • *Enter quickly.*

81. What is meant by cook off a grenade?
    Pull the pin and allow the safety lever to fly; hold for two seconds and then toss.

82. Name some things that are considered fighting positions in urban combat.
    • *The corner of a building.*
    • *Walls.*
    • *Windows.*
    • *Roof peaks.*
    • *Loopholes.*

83. What are the six fundamentals of tracking?
    (1) Displacement.
    (2) Staining.
    (3) Weathering.
    (4) Littering.
    (5) Camouflage.
    (6) Interpretation and/or immediate use intelligence.

84. What are some things the analysis of a track can tell you?
    • *The direction and rate of movement of a party.*
    • *The number of persons in a party.*

- *Whether or not heavy loads are carried.*
- *The sex of the members of a party.*
- *Whether the members of a party know they are being followed.*

85. How can you tell how many people walked on a trail?
By use of the box method. Mark off a 30- to 36-inch section of a trail. Count the number of footprints in the box and divide by two.

86. In tracking, what are some examples of displacement?
- *Turned over rocks or sticks.*
- *Crushed or disturbed vegetation.*
- *Slip marks and water-filled footprints on stream banks.*

87. What is assault fire?
Assault fire is walking rapidly and firing your rifle from the under-arm or quick-fire position, and then stopping momentarily to take well-aimed directed shots when definite targets appear.

88. What are some silent weapons?
The knife or bayonet, garrote, and a club.

89. What defines the Warrior Ethos?
It is defined by four lines from the Soldiers' Creed:
(1) I will always place the mission first.
(2) I will never accept defeat.
(3) I will never quit.
(4) I will never leave a fallen comrade.

90. The battlefields of the Global War on Terrorism and anticipated future battlefields are expected to be described as what?
- *Asymmetrical.*
- *Violent.*
- *Unpredictable.*
- *Multidimensional.*

91. What skills do you need for fighting in an urban area?
- *Moving.*
- *Entering buildings.*
- *Clearing rooms.*
- *Selecting and using fighting positions.*

92. Name nine Warrior Drills.
    (1) React to contact.
    (2) React to a near ambush.
    (3) React to a far ambush.
    (4) React to indirect fire.
    (5) React to a chemical attack.
    (6) Break contact.
    (7) Dismount a vehicle.
    (8) Evacuate wounded personnel from a vehicle.
    (9) Establish security at the halt.

93. What are the key movement techniques?
    * *Avoid open areas.*
    * *Avoid silhouetting yourself.*
    * *Select the next covered position before moving.*

94. What basic information goes on a captured document tag?
    * *Nationality of the capturing force.*
    * *Date-time group.*
    * *Place captured.*
    * *Identity of source.*
    * *Circumstances of capture.*
    * *Description of document or weapon.*

95. What factors affect the estimating of ranges?
    * *Nature of the object.*
    * *Nature of the terrain.*
    * *Light conditions.*

96. The U.S. National Policy on Antipersonnel Land Mines forbids what?
    * *The use of non self-destructing antipersonnel mines except in an area like the DMZ of South Korea.*
    * *The use of standard or improvised explosive devices as booby traps.*
    * *Training with M14 or M16 mines except for units deploying to South Korea or training for mine removal (EOD).*

97. The U.S. National Policy on Antipersonnel Land Mines allows what?
    - *The use of antivehicle mines.*
    - *The use of the M18 Claymore mine in the command detonated mode.*
    - *The use of mixed minefields using self-destructing anti-personnel mines.*

98. What is the M-131 Modular Pack Mine System?
    The MOPMS is a man-portable antitank and antipersonnel system containing a mix of M78 antiarmor and M77 antipersonnel mines.

99. How much does the M-131 MOPMS weigh?
    165 pounds.

100. How is the M-131 MOPMS activated?
    By hard wire or radio communication.

101. After the mines in the M-131 are dispersed can they be recovered?
    No.

102. What is an IED?
    A nonstandard explosive device that targets soldiers and civilians.

103. What are the three basic types of IEDs?
    (1) Timed explosive devices.
    (2) Impact-detonated devices.
    (3) Vehicle borne bombs.

104. What do you watch for when searching for IEDs?
    - *Wires.*
    - *Antennas.*
    - *Detcord.*
    - *Parts of exposed ordnance.*

105. What actions are to be taken when a suspected IED is found?
    - *Maintain 360-degree security.*
    - *Move away to a minimum of 300 meters.*
    - *Scan area for additional IEDs.*
    - *Cordon off the area to prevent civilian casualties.*
    - *Report the situation to the next highest command.*

106. What are the five methods of determining direction to a target?
    (1) Estimating.
    (2) Scaling from a map.
    (3) Using a compass.
    (4) Measuring from a reference point.
    (5) Using other devices.

107. What are some methods of determining the range to a target?
    • *Football field method.*
    • *Flash-to-bang method.*
    • *Binocular-reticule/mill-relation method.*
    • *Recognition and appearance of an object method.*

108. Current and future operational environments are expected to have what characteristics?
    • *Constant high intensity close combat.*
    • *No rear areas, no sanctuary.*
    • *Information operations effects down to the tactical level.*
    • *Constantly changing rules of engagement and tactics.*
    • *Combat and non-combat roles blurred.*
    • *Extreme stress with leader fatigue.*

109. What are the problem-solving steps?
    • *Problem definition*
    • *Information gathering*
    • *Course of action development*
    • *Course of action analysis*
    • *Course of action comparison*
    • *Decision*
    • *Execution and assessment*

110. What are the five forms of maneuver?
    (1) Envelopment
    (2) Turning movement
    (3) Infiltration
    (4) Penetration
    (5) Frontal attack

Battle-Focused Training

REFERENCES: FM 7-0, Training for Full-Spectrum Operations;
ADP 7-0, Training Units and Developing Leaders;
ADRP 7-0, Training Units and Developing Leaders

1. What is considered the cornerstone of readiness?
   Training.

2. What is training?
   The means to achieve the tactical and technical proficiency that soldiers, leaders, and units must have to enable them to accomplish their mission.

3. What type of training is most preferred and most effective?
   Performance-oriented training (hands-on).

4. If a soldier receives a NO-GO during performance-oriented training, what should the trainer do next?
   Tell him what he did wrong, follow up with more practice, and then retest.

5. Who has responsibility for collective and individual soldier training?
   Officers are responsible for collective training, and NCOs are responsible for individual soldier training.

6. What types of training can an NCO impose on his subordinates?
   Motivational, reinforcement, and remedial training.

7. Who is the primary trainer in a unit?
   The commander.

8. Who said, "The best form of welfare for the troops is first class training, for this saves unnecessary casualties?"
   Field Marshal Erwin Rommel.

9. Name five of the eleven principles of training.
   (1) NCOs train individuals, crews, and small teams.
   (2) Train as you will fight.
   (3) Train to standards.
   (4) Train while operating.

(5) Train fundamentals first.
(6) Train to sustain proficiency.
(7) Understand the operational environment.
(8) Train to maintain.
(9) Make commanders the primary trainers.
(10) Train to develop adaptability.
(11) Conduct multi-echelon and concurrent training.

10. What starts the training planning process?
An assessment.

11. Specify the two types of training assessment.
Testing and the Army Training and Evaluation Program (ARTEP).

12. What are the three stages of training?
Initial, refresher, and sustainment.

13. What does well-structured training contain?
A mixture of initial and sustainment training.

14. What are the three methods used to present training to soldiers?
Lecture, conference, and demonstration.

15. What are the steps on the training ladder?
Collective training, leader training, and individual training.

16. Where would you go to find appropriate Army doctrine?
   • *Field manuals.*
   • *Training circulars.*
   • *Mission training plans.*
   • *Drill books.*
   • *Soldier's manuals.*
   • *Army regulations.*

17. What manual is considered the bible for collective training?
The Army Training and Evaluation Program (ARTEP) manual.

18. What manual is considered the bible for individual training?
Soldier's Manual.

19. State the four steps in preparing individual training.
    (1) Gather materials.
    (2) Announce training.
    (3) Rehearse training.
    (4) Revise training outline.

20. What are the three parts of a training objective?
    Task, Condition, and Standard.

21. What are the three phases in the leader development program?
    (1) Reception and integration.
    (2) Basic skill development.
    (3) Advanced development and sustainment.

22. What are the keys to successful performance training?
    - *Focus on fundamentals.*
    - *Live-fire exercises.*
    - *Night and adverse weather training.*
    - *Drills.*
    - *Lane training.*
    - *Incorporate competition.*
    - *Conduct post-training checks.*

23. What is the objective of all Army training?
    Unit readiness.

24. What does the ideal command climate promote?
    - *Learning.*
    - *Honest mistakes.*
    - *Open communication.*
    - *Discipline.*

25. What characteristic can be expected in future operational environments?
    Constant, high intensity, close combat
    No rear areas; no sanctuary
    Constant changing of rules of engagement and tactics
    Combatant and non-combatant roles blurred
    Extreme stress and leader fatigue

26. What is the number one principle of peacetime training?
Replicate battlefield conditions.

27. What does the acronym METL stand for?
Mission Essential Task List.

28. A unit's Mission Essential Task List is based on what?
Its wartime mission.

29. Name the two primary inputs to METL development.
War plans and external directives.

30. What does the acronym METT-TC stand for?
—Mission.
—Enemy.
—Terrain.
—Troops.
—Time available.
—Civilian considerations.

31. Name five of the seven battlefield operating systems.
(1) Intelligence.
(2) Maneuver.
(3) Fire support.
(4) Mobility, countermobility, survivability.
(5) Air defense.
(6) Combat service support.
(7) Command and control.

32. What is a combined training exercise?
A training exercise that is conducted by military forces of more than one nation.

33. The greatest combat power is a result of what?
Leaders synchronizing combat support and combat service support systems to complement and reinforce one another.

34. What is risk assessment?
The thought process of making operations safer without compromising the mission.

35. Historically speaking, what is the cause of the majority of casualties in combat?
    Accidents.

36. What methods are used to assess soldier, leader, and unit proficiency?
    Evaluations.

37. What are the two types of evaluations?
    Formal and informal.

38. Identify the two types of after action reviews.
    Formal and informal.

39. Enumerate the four parts of an after action review (AAR).
    (1) Establish what happened.
    (2) Determine what was right or wrong with what happened.
    (3) Determine how the task should be done differently the next time.
    (4) Perform the task again.

40. What does challenging training result in?
    (1) Aggressive and well-trained soldiers.
    (2) Higher levels of competence and confidence.
    (3) Greater loyalty and dedication.
    (4) Inspired excellence through initiative, enthusiasm, and eagerness to learn.

41. Each soldier is responsible for performing individual tasks. Where do these tasks come from?
    They are tasks the first line supervisor identifies on the unit's mission-essential task list (METL).

42. What are the three training domains?
    Institutional, operational, and self-development.

43. What does the acronym TADSS stand for?
    Training Aids, Devices, Simulators, and Simulations.

44. What must training be to be efficient?
    Relevant, rigorous, realistic, challenging, and properly resourced.

45. What are the principles of leader development?
    Lead by example
    Develop subordinate leaders
    Create a learning environment
    Train leaders in the art and science of mission command
    Train to develop adaptive leaders
    Train leaders to think critically and creatively
    Train leaders to know subordinates and their families.

46. What is the Army Training Network?
    An Army online portal that helps units and leaders plan and conduct unit training.

Camouflage, Concealment, and Decoys

REFERENCES: FM 20-3 Camouflage, Concealment, and Decoys;
FM 3-21.75, The Warrior Ethos and Soldier Combat Skills

1. What is the difference between cover and concealment?
   Cover offers protection from enemy fire; concealment offers protection against enemy observation.

2. Why should your fighting position have overhead cover?
   For protection from mortar fire and fragments from other indirect-fire weapons.

3. Define "direct observation."
   When the observer sees the subject physically with his eyes.

4. Define "indirect observation."
   When the observer sees an image of the subject and not the subject itself, such as with radar, infrared equipment, or television.

5. Give some examples of natural concealment.
   Bushes, tall grass, and shadows.

6. What is camouflage?
   The measures taken to conceal yourself, your equipment, and your position.

7. Who is responsible for overall unit camouflage?
   The unit commander.

8. Who is responsible for the individual soldier's concealment?
   The individual soldier.

9. Name the three principles employed to eliminate the factors of recognition.
   Siting, discipline, and construction.

10. State the three methods of concealing installations and activities.
    Hiding, blending, and disguising.

11. List the five camouflage measures.
    (1) Siting and dispersion.
    (2) Use of natural materials.
    (3) Pattern painting.
    (4) Nets.
    (5) Digging in.

12. If you camouflage properly, how close should the enemy be able to approach before detecting your position?
    Approximately 35 meters—within hand grenade range.

13. List items you can use to camouflage the shiny parts of your weapon.
    • *Burlap.*
    • *Sandbags.*
    • *Cloth strips.*
    • *Camouflage sticks.*

14. Name the main items you must consider when starting to camouflage.
    • *Shape.*
    • *Shine.*
    • *Outline and shadows.*
    • *Color.*
    • *Position.*
    • *Movement.*
    • *Dispersion.*

15. What are the shiny portions of the face?
    • *Cheekbones.*
    • *Nose.*
    • *Forehead.*
    • *Chin.*

16. Must a dark-skinned person camouflage his skin? Why?
    Yes, because of body oils.

17. Mud may be used in an emergency for camouflage. What are the disadvantages in using it?
    It changes color as it dries out, it peels off, and it may contain harmful bacteria.

18. What is an oblique area photo?
    A photo that is taken at an angle from the vertical (the camera is not parallel to the horizon).

19. How can you distinguish a high-oblique area photo?
    The apparent horizon can be seen.

20. When was camouflage first used on a large scale?
    World War I.

21. What are the two basic types of cover?
    Natural and man-made.

22. Name some types of natural cover.
    Logs, trees, stumps, ravines, and hollows.

23. Name some types of man-made cover.
    Fighting positions, trenches, walls, rubble, and craters.

24. What should you do before starting to apply camouflage?
    Study the vegetation and terrain.

25. What are the shadow portions of the face?
    Around eyes, under nose, and under chin.

26. What are the six principles of camouflage?
    - *Light.*
    - *Heat.*
    - *Noise.*
    - *Spoil.*
    - *Trash.*
    - *Movement.*

27. What is the purpose of camouflage, cover, and concealment?
    To deny observation of your unit, equipment, and position and to deny this information to the enemy.

Code of Conduct

REFERENCE: AR 350-30, Code of Conduct, Survival, Evasion,
Resistance and Escape (SERE) Training

1. Under which presidential administration was the Code of Conduct
   established?
   The Eisenhower administration, on 17 August 1955.

2. What president amended the Code of Conduct in 1988?
   President Ronald Reagan.

3. What happened to the Code of Conduct in 1988?
   It was amended with language that was gender neutral.

4. What document supports the Code of Conduct?
   *Manual for Courts-Martial under Uniform Code of Military
   Justice.*

5. Why was the Code developed?
   *To provide a form of mental defense for U.S. POWs to use to resist
   enemy POW management practices.*

6. What is the Code of Conduct?
   *The Code is a guide for soldier conduct if he or she must evade
   capture, resist as a captive, and attempt to escape.*

7. What does the Code of Conduct mean, expressed in your own
   words?
   *(Example) It is a written law by Executive Order that governs
   soldiers' actions and conduct during time of war should they
   become captured or prisoners of war.*

8. Does the Code also apply to soldiers held by terrorists in peacetime?
   Yes.

9. What is the first sentence of the Code?
   *"I am an American, fighting in the forces that guard my country
   and our way of life."*

10. How many articles are in the Code of Conduct?
    Six.

11. Which article of the Code pertains to escape and evasion?
    Article 3.

12. Should you become a prisoner of war, what information are you required to give when questioned?
    Name, rank, service number, and date of birth.

13. What is your priority if captured?
    Attempt to escape.

14. What are the four actions a soldier must not do as a POW?
    (1) Make oral or written confessions.
    (2) Answer questions.
    (3) Provide personal histories.
    (4) Appeal for surrender.

15. As an individual, may any member of the armed forces voluntarily surrender?
    No.

16. What does the acronym SERE stand for?
    Survival, evasion, resistance, and escape.

17. What are the three personal values that will sustain a soldier in surviving captivity?
    Courage, dedication, and motivation.

18. What does the United States promise you under the code?
    (1) To keep faith in me and to stand by me as I fight for its defense;
    (2) To care for my family and dependents; and
    (3) To use every practical means to contact, support, and gain release for me and all other prisoners of war.

Combat Stress

REFERENCE: FM 6-22.5, Combat Stress

1. What is combat stress?
   The mental, emotional, or physical tension, strain, or distress resulting from exposure to combat and combat-related conditions.

2. Who is the key to building and maintaining high unit morale and peak efficiency?
   The small unit leader.

3. What is another name for combat stress?
   Combat fatigue and Post-Traumatic Stress Disorder (PTSD).

4. What is the basic rule for combat stress recognition?
   Know your troops and be alert for any sudden, persistent, or progressive change in their behavior that threatens the functioning and safety of your unit.

5. What are some causes for combat stress?
   - *Fear and anxiety.*
   - *Heavy physical work.*
   - *Sleep loss.*
   - *Dehydration.*
   - *Poor nutrition.*
   - *Severe noise and vibrations.*
   - *Exposure to heat, cold, and wetness.*
   - *Poor hygiene facilities.*
   - *Exposure to disease.*
   - *Toxic fumes or substances.*

6. What are some physical signs of mild stress?
   - *Trembling.*
   - *Jumpiness.*
   - *Cold sweats and dry mouth.*
   - *Insomnia.*
   - *Pounding heart.*
   - *Dizziness.*
   - *Nausea, vomiting, diarrhea.*
   - *Fatigue.*
   - *"Thousand yard" stare.*
   - *Difficulty thinking, speaking, or communicating.*

7. What must you be aware of with a soldier displaying signs of stress?
   Suicide.

8. What are some emotional signs of mild stress?
   • *Anxiety/indecisiveness.*
   • *Irritability/complaining.*
   • *Forgetfulness.*
   • *Nightmares.*
   • *Easily startled.*
   • *Tears/crying.*
   • *Anger/loss of confidence in self or unit.*

9. What are some physical signs of severe stress?
   • *Constantly moving around.*
   • *Flinching or ducking at sudden sounds or movement.*
   • *Shakes or tremors.*
   • *Can not use a part of their body for no apparent physical reason.*
   • *Inability to see, hear, or feel.*
   • *Physical exhaustion.*
   • *Freezes under fire.*
   • *Stares vacantly, staggers or sways when standing.*
   • *Panics, runs under fire.*

10. What are some emotional signs of severe stress?
    • *Talks rapidly or inappropriately.*
    • *Argumentative.*
    • *Memory loss.*
    • *Stutters or can't speak at all.*
    • *Sees or hears things that do not exist.*
    • *Rapid emotional shifts.*
    • *Socially withdrawn.*
    • *Apathetic.*
    • *Hysterical outbursts.*
    • *Frantic or strange behavior.*

Customs and Laws of War

REFERENCE: FM 27-60, The Law of Land Warfare

1. What are the purposes of laws of war?
   To protect both combatants and noncombatants from unnecessary suffering, to safeguard human rights, and to restore peace.

2. During war, which personnel are normally considered noncombatants?
   - *Civilians.*
   - *Medical personnel.*
   - *Chaplains.*
   - *Soldiers who surrender or are sick or wounded.*

3. A U.S. treaty drafted in 1785 with the Kingdom of Prussia is considered the first agreement concerning wartime. By whom was it drafted and for what reason?
   It was drafted by Benjamin Franklin, John Adams, and Thomas Jefferson to improve the treatment of POWs.

4. Where was the Geneva Convention written?
   Geneva, Switzerland.

5. How many nations were initially represented in the drafting of the first Geneva Convention?
   Sixty-one nations.

6. How many articles in the Geneva Convention relate to the treatment of POWs?
   143 articles.

7. What do the Geneva Conventions recognize as the main duty of a POW?
   To try to escape and/or aid others in escaping.

8. What documents may a POW keep upon capture?
   Identity cards and personal papers.

9. Enumerate the rights of a POW.
   - *To have enough food to stay in good health.*
   - *To have sanitary housing and clothing.*
   - *To have facilities for personal hygiene.*
   - *To practice your religious faith.*
   - *To keep personal property.*
   - *To send and receive mail.*
   - *To receive packages containing food, clothing, educational, and recreational materials.*
   - *To select a fellow POW to represent you.*
   - *To receive humane treatment.*
   - *To have a copy of the Geneva Convention posted where you can read it.*
   - *To have a copy of the camp regulations posted.*

10. What type of work can be imposed upon enlisted POWs?
    Work required to maintain them in a good state of physical and mental health, or work to improve their environment, as long as the work has no military character or purpose.

11. What type of work can an NCO-POW be forced to do?
    Supervisory work.

12. Your squad has captured some enemy soldiers. These POWs have weapons, gas masks, and personal property in their possession. What can you have your soldiers take from them?
    Only their weapons.

13. As a minimum, what data must be included on an enemy POW capture tag?
    - *Date and time of capture.*
    - *Capturing unit.*
    - *Place of capture.*
    - *Circumstances of capture.*

14. What action can be taken against a soldier if he or she violates the Geneva Conventions?
    Court martial under UCMJ.

Drill and Ceremonies

REFERENCE: TC 3-21.10, Drill and Ceremonies

1. In marching, what is meant by a drill?
   Movement by a unit in an orderly manner from one place to another with the movements executed in unison with precision.

2. What is the historical purpose of drills?
   The conquest of fear, achieved through the loss of individuality and the unification of a group under obedience to orders.

3. What is accomplished by practicing dismounted drills?
   * *Teamwork.*
   * *Confidence.*
   * *Pride.*
   * *Alertness.*
   * *Attention to detail.*
   * *Esprit de corps.*
   * *Discipline.*

4. What was the "Blue Book?"
   A drill manual used by Baron Von Steuben to train the colonial army. It was the predecessor of FM 3-21.5.

5. What are the three methods used to teach drill?
   Step by step, by the numbers, and talk-through.

6. When marching, who is the only person in a platoon that is never out of step?
   The platoon guide.

7. What is cadence?
   The uniform step and rhythm used in marching, or the number of steps or counts per minute at which a movement is executed.

8. How is a marching step measured?
   From heel to heel.

9. What is the length of the following steps: forward, backward, half step, double time, and left or right?
   Forward and double time are 30 inches. Half step, backward, and left or right step are 15 inches.

10. How many steps per minute is the cadence for double time?
    180 steps.

11. How many steps per minute is the cadence for quick time?
    120 steps.

12. Name the four movements in marching that require a 15-inch step.
    (1) Half step.
    (2) Left step.
    (3) Right step.
    (4) Backward march.

13. What is the proper length of arm swing when marching?
    Nine inches to the front and six inches to the rear.

14. Most drill commands have two parts. Specify them.
    The preparatory command and the command of execution.

15. What are the first two commands the first sergeant gives when he forms the unit?
    Fall in and Receive the report.

16. What command is given to revoke a preparatory command?
    As you were.

17. Can a command be changed after the command of execution has been given?
    No.

18. What is a supplementary command?
    An order given by a subordinate leader that reinforces and complements a commander's order.

19. If you were marching a squad, when would you give the command "Squad, halt?"
    When either foot strikes the ground.

20. What are some drill commands in which the preparatory command and the command of execution are combined?
    Fall in, At ease, and Rest.

21. When under arms, at what position would you "Fall In?"
    Order arms.

22. After weapons have been issued to a unit and all soldiers have fallen in with their weapons, what is the next command that you should give?
    Inspection arms.

23. When performing drills with the M16, what is the only command that can be given from inspection arms?
    Ready, Port, Arms.

24. From what positions may a rifle salute be executed?
    Right and left shoulder arms, trail arms, and order arms.

25. What is meant by the command "Cover?"
    Align yourself directly behind the soldier to your immediate front while maintaining the proper distance.

26. To align your squad at close interval, what two commands would you give?
    At close interval dress right, dress, and Ready front.

27. On what foot would you give the command "Mark time, March?"
    On either foot.

28. When starting from the halt, all steps begin with the left foot except one. What movement is it?
    Right step, March.

29. Identify the four rest positions that can be given at the halt.
    (1) Parade rest.
    (2) Stand at ease.
    (3) At ease.
    (4) Rest.

30. What is the only position from which the command "Parade rest" is given?
    From the position of attention.

31. Describe the position of "rest."
You may move, talk, smoke, or drink unless otherwise specified. You must remain standing with your right foot in place.

32. What command is given to reverse the direction of march?
Rear march.

33. When is the command "Right turn" or "Left turn, march" used?
When marching elements of more than four columns abreast.

34. When executing "Rear march," on which foot do you pivot?
On both feet, turning 180 degrees to the right.

35. Prior to giving the command "Double time," what command must be given if the unit is armed with rifles?
Port arms or Sling arms.

36. What commands in the manual of arms are not executed in cadence?
Fix bayonets and Unfix bayonets.

37. What is an interval?
The lateral space between elements in a formation.

38. When the platoon is formed on line, the post of the platoon leader is how many steps in front of the platoon?
Six steps and centered.

39. What are the two prescribed formations for a platoon?
A platoon on line and a platoon in column.

40. How many steps does each of the four ranks take when a platoon is given open ranks?
On the command "March": First rank takes two steps forward, second rank takes one step forward, third rank stands fast, and fourth rank takes two steps backward.

41. When a platoon is armed, the platoon sergeant dismisses the platoon by what commands?
    - *Inspection arms.*
    - *Ready.*
    - *Port arms.*
    - *Order arms.*
    - *Dismissed.*

42. How many steps should separate platoons when a company is formed?
    Five steps.

43. What is a muster formation?
    A formation to call roll to determine accountability of personnel.

44. What is a rank?
    A single line of personnel standing side by side.

45. What is a file?
    A single line of personnel standing one behind the other.

46. What is meant by a ceremony?
    Formation and movement in which a number of troops execute movements as a drill with the primary purpose to render honors, preserve tradition, and stimulate esprit de corps.

47. What are the two bugle calls played at retreat? Name them in the order they are played.
    "Retreat" and "To the Color."

48. Reveille is a ceremony performed in the morning. Who determines when reveille will be sounded?
    The installation commander.

49. How many soldiers are in the firing party for a ceremonial firing squad?
    Eight—seven firers and one noncommissioned officer in charge (NCOIC).

50. List the recommended components of an enlisted burial escort.
    - *Bugler.*
    - *Six pallbearers.*
    - *Seven-man firing party.*
    - *NCOIC.*

51. Who is responsible for the training and appearance of the color guard?
    The command sergeant major (CSM).

52. When does the national color render a salute?
    Never.

53. How far in front of the honor or color company do the colors halt?
    Ten steps.

54. While passing the colors or while the colors are passing you, when is the hand salute rendered?
    Six paces before and six paces after.

55. How far in front of the color company do the colors halt?
    Ten steps.

56. Does a color guard ever execute rear march?
    No.

57. When at Route step, March, what command brings you back to attention?
    Quick time, March.

58. Facing movements are executed at what position while under arms?
    Order arms or Sling arms.

59. What was the Blue Book's official name?
    The Regulations for the Order and Discipline of the Troops of the United States.

Enlisted Promotions

REFERENCE: AR 600-8-19, Enlisted Promotions and Reductions

1. What AR covers enlisted promotions?
   AR 600-8-19

2. What are decentralized promotions?
   Local promotions of Specialist and below.

3. What are semi-centralized promotions?
   Promotions to Sergeant and Staff Sergeant.

4. What are centralized promotions?
   Promotions to Sergeant First Class, Master Sergeant, and Sergeant Major.

5. How are First Sergeant and Command Sergeant Major positions filled?
   By appointment.

6. What is the purpose of the Enlisted Promotion System?
   The purpose is to fill authorized enlisted spaces with the best qualified and to provide career progression and recognition for the best-qualified soldiers.

Field Hygiene and Sanitation

REFERENCES: FM 4-25.12, Unit Field Sanitation Team;
FM 21-10, Field Hygiene and Sanitation

1. What is field sanitation?
The use of measures to create and maintain healthful environmental conditions; these include safeguarding food, safeguarding water, and controlling disease-bearing insects and rodents.

2. Name the five Fs of field sanitation.
(1) Feces.
(2) Fingers.
(3) Flies.
(4) Foods.
(5) Fluids.

3. What are the four types of waste?
(1) Human.
(2) Liquid.
(3) Garbage.
(4) Rubbish.

4. Specify the two critical temperatures for maintaining food in the field.
Below 45 degrees Fahrenheit (F) for cold foods and above 140 degrees F for hot foods.

5. If hot soapy water is not available to clean your mess kit, what can you use in its place?
Sand and gravel.

6. How far should garbage pits be located from bivouac mess areas?
30 yards.

7. For what is a soakage pit used?
To prevent the accumulation of liquid waste water.

8. Name at least three devices most generally used for disposal of human waste while in the field.
(1) Straddle-trench latrines.
(2) Deep-pit latrines.

(3) Burn-out latrines.
(4) Pail latrines.
(5) Urine soakage pits.

9. A latrine is usually closed when it is how full?
   When it is within one foot from the surface.

10. A latrine can be no closer than what distance to a unit mess site?
    100 yards.

11. A latrine can be no closer than what distance to a water source?
    30 yards.

12. What type of latrine is specifically used in an area with a high water table?
    Burn-out or pail latrine.

13. If you are in a temporary bivouac for one to three days, what type of latrine will you normally use?
    A straddle trench.

14. How many straddle-trench latrines would be needed for one hundred male soldiers?
    Two.

15. How many straddle-trench latrines would be needed for one hundred female soldiers?
    Three.

16. How many four-hole latrines would be required for one hundred men?
    Two.

17. What type of latrine is normally used by soldiers on the march?
    A cat hole or individual waste collection bag.

18. How deep is a cat hole?
    Approximately one foot.

19. List the three rules for water discipline while in the field.
    Drink only from approved sources, conserve water, and don't contaminate your sources.

20. Which is absorbed fastest by the body—warm, cool, or cold water?
    Cool water (at 50 to 55 degrees Fahrenheit).

21. What is the water requirement per person per day in a temperate zone?
    Four gallons or more.

22. List the six sources of water.
    (1) Surface.
    (2) Ground.
    (3) Rain.
    (4) Ice.
    (5) Snow.
    (6) Sea water.

23. What is potable water?
    Water that is safe to drink.

24. Who gives the final approval on whether water is fit for human consumption?
    Medical personnel.

25. For how long must water be boiled before it is safe for drinking?
    At least fifteen minutes.

26. What chemicals are normally used to purify water while in the field?
    Iodine tablets, chlorine ampules, household bleach.

27. What is the common name for a water purification bag?
    A Lyster bag.

28. How many gallons of water will a Lyster bag hold?
    Up to thirty-six gallons.

29. What is the greatest preventive measure for disease?
    Cleanliness.

30. What are the two main measures a soldier can take to keep from getting sick?
    Keep the body clean and keep the shot record up-to-date.

31. What measures do soldiers need to take to ensure maximum resistance to disease?
Maintain good mental and physical health, practice personal hygiene, and get necessary immunizations.

32. In enforcing sleep discipline, how many hours of sleep should a soldier get in each twenty-four-hour period?
Six to eight under normal conditions, three to four under continuous operations.

33. Why is each soldier issued two pairs of boots?
To rotate them and air them out.

34. What is artificial immunity?
Resistance to infection and disease acquired from vaccinations that stimulate the body to produce antibodies, or from immunizing serums already containing the desired antibodies.

35. Identify the five communicable disease groups as classified by the Army.
(1) Respiratory.
(2) Intestinal.
(3) Insect-borne.
(4) Venereal.
(5) Miscellaneous.

36. Name the three parts of the disease transmission chain.
Reservoir (source), vehicle, and susceptible person.

37. What is the meaning of the term vector as related to sanitation?
Any carrier of a disease.

38. Historically, what are the four major medical threats to field soldiers?
(1) Heat injury.
(2) Cold injury.
(3) Diarrheal diseases.
(4) Arthropod- (insect-) transmitted diseases.

39. Name some individual preventive medicine measures (PMMs) to prevent heat injury.
    - *Drink plenty of water.*
    - *Use work/rest cycles.*
    - *Eat all meals for energy and to replace salt.*
    - *Recognize the risks in mission-oriented protective posture (MOPP) and body armor.*
    - *Modify uniforms.*

40. What is a major sign of not drinking enough water?
    Dark yellow urine.

41. What is the better indicator of dehydration, thirst or urine color?
    Urine color.

42. Mention the dangers or conditions that a soldier should guard against while in the field during cold weather.
    - *Trench foot.*
    - *Immersion foot.*
    - *Frostbite.*
    - *Snow blindness.*
    - *Carbon monoxide poisoning.*

43. What factors affect cold-weather injuries?
    - *Age.*
    - *Rank.*
    - *Previous exposure.*
    - *Fatigue.*
    - *Discipline.*
    - *Nutrition.*
    - *Activity.*
    - *Geographic origin.*
    - *Use of drugs.*

44. What are the three major diseases carried by mosquitoes?
    Malaria, yellow fever, and encephalitis.

45. Name three germs that flies may carry.
    Typhoid, cholera, and dysentery.

46. Name the three most common poison plants.
    Poison ivy, poison oak, and poison sumac.

47. Name some toxic chemical threats other than those that are nuclear, biological, and chemical (NBC).
    Carbon monoxide; bore/gun gases; and solvents, greases, or oils.

48. Itemize the cans in a mess kit laundry.
    Four: a scrap can, hot soapy water wash can, and two rinse cans of hot boiling water.

49. How hot should the soapy water in a mess kit laundry be kept?
    120 to 140 degrees F.

50. Historically, what percentage of casualties have been from disease and noncombat injuries?
    80 percent.

51. According to FM 21-10, how often must soldiers shower in the field?
    All personnel must bathe at least once a week and have a clean change of clothing.

52. What kind of undergarments should be worn in the field, and why?
    Cotton. Cotton will breathe and dry out.

53. Leaders and field sanitation teams must do what in the field?
    - *Inspect water containers and trailers.*
    - *Disinfect unit water supplies.*
    - *Check unit water supply for chlorine.*
    - *Inspect unit field food service operations.*
    - *Control arthropods, rodents, and other animals in unit areas.*
    - *Train unit personnel in the use of individual preventive medical measures (PMM).*
    - *Monitor status of PMM in unit.*
    - *Assist in the selection of a unit bivouac site.*
    - *Supervise the construction of field sanitation devices.*
    - *Monitor unit personnel in the application of individual PMM.*

Field Mess Operations

REFERENCES: FM 10-23, Basic Doctrine for Army Field Feeding and Class 1 Operations; FM 8-34, Food Sanitation for the Supervisor

1. How hot should the soapy water in a mess kit wash line be?
   120 to 140 degrees F.

2. A mess kit laundry in a field kitchen should contain four 32-gallon steel cans. What are their uses?
   (1) Waste.
   (2) Wash.
   (3) Rinse.
   (4) Sanitize.

3. The gasoline storage area for a field mess should be at least how far from the kitchen?
   50 feet (15 meters).

4. A properly prepared insulated food container will keep hot food warm for how long?
   Three to five hours.

5. What are the ideal characteristics of a soldier who prepares or serves food?
   - *A clean uniform.*
   - *Clean hands and fingernails.*
   - *A head covering, be clean shaven (if male).*
   - *Be free of illness or infection.*
   - *Wear no jewelry except a wedding band.*

6. Identify the types of rations you can expect to eat while in the field.
   - *Fresh, perishable foods (A-rations).*
   - *Canned and dehydrated foods (B-rations).*
   - *Prepared table pans or trays (T-rations).*
   - *Meals ready to eat (MREs).*

7. What are the three sources or types of food contamination?
    (1) Biological hazards—usually by improper food handling.
    (2) Chemical hazards—usually by accidental occurrence.
    (3) Physical hazards—usually foreign particles through inadequate food protection.

8. What are the five unsafe food handling practices?
    (1) Failure to refrigerate cold foods below 40 degrees F.
    (2) Failure to maintain hot foods at 140 degrees F or above.
    (3) Not protecting food from cross contamination.
    (4) Improper food transportation or storage.
    (5) Improper procedures and practices of food handlers.

First Aid

REFERENCES: FM 4-25.11, First Aid;
TRADOC PAM 600-4, IET Soldier's Handbook;
FM 3-21.75, The Warrior Ethos and Soldier Combat Skills

1. What is first aid?
   The immediate care given to the sick, injured, or wounded by non-medical personnel until professional medical treatment can be obtained.

2. What is the object of first aid?
   To stop bleeding, overcome shock, relieve pain, and prevent infection.

3. What is self aid?
   Emergency treatment you apply to yourself.

4. Why is medical training so important?
   A medical emergency may occur where medical personnel are not immediately available.

5. Itemize the ten steps in evaluating a casualty.
   - *Check for responsiveness.*
   - *Check for breathing.*
   - *Check for bleeding.*
   - *Check for shock.*
   - *Check for fractures.*
   - *Check for burns.*
   - *Check for head injury.*
   - *Seek professional medical aid.*
   - *Perform all necessary steps in sequence.*
   - *Identify all wounds and/or conditions.*

6. What are the four lifesaving steps?
   (1) Open the airway and restore breathing and heartbeat.
   (2) Stop the bleeding.
   (3) Dress the wound.
   (4) Control for shock.

7. Describe two methods to open an airway.
   Head-tilt, chin-lift method and jaw-thrust method.

8. Name two types of artificial respiration.
Mouth-to-mouth and the back-pressure, arm-lift method.

9. When giving mouth-to-mouth resuscitation, what should you do if the casualty resumes breathing?
Watch the casualty closely to maintain an open airway and check for other injuries.

10. What does a bulging stomach in a mouth-to-mouth patient indicate?
Air is entering the stomach; reposition the head and continue.

11. What assistance is given to restore heartbeat?
Cardiopulmonary resuscitation.

12. For what does CPR stand?
Cardiopulmonary resuscitation.

13. When performing CPR, what are the ratios of breaths to chest compressions?
With one person, it is two to fifteen. With two people, it is one to five.

14. What does the term hemorrhage mean?
Heavy bleeding.

15. Identify the three types of bleeding.
Arterial, venous, and capillary.

16. How can you recognize arterial bleeding?
By spurts of bright red blood.

17. Name the ways to control bleeding.
   - *Using pressure bandage.*
   - *Using pressure points.*
   - *Elevating the bleeding area.*
   - *Applying a tourniquet.*

18. How many pressure points are there on one side of the body?
Six.

19. How will you know you are in the right location when applying digital pressure to control bleeding?
    You'll feel a pulse.

20. How long is direct, manual pressure applied in order to control bleeding?
    Five to ten minutes.

21. How high should an injured limb be elevated above the level of the heart to control bleeding?
    Two to four inches.

22. Cite a situation in which you would use a tourniquet on a casualty.
    When all other methods have failed to control bleeding.

23. After placing a tourniquet on a casualty, what do you do?
    Mark the forehead with a T and note the time of application.

24. When would you remove a tourniquet?
    Never. It should be removed only by medical personnel.

25. How do you take the carotid pulse?
    Place the first two fingers of your hand beside the casualty's Adam's apple.

26. What blood type is considered a universal-donor type?
    Type O.

27. Tick off the steps in treating a sucking chest wound.
    - *Seal wound with airtight material when casualty exhales.*
    - *Secure in place with tape, bandage, or cravat.*
    - *Cover completely with dressing.*
    - *Have casualty lie on injured side.*
    - *Get medical help as soon as possible.*

28. What is the main indication of a sucking chest wound?
    Frothy fluid bursting from the wound with each breath.

29. Itemize the symptoms or signs of a casualty in shock.
    - *Clammy skin.*
    - *Thirst.*

- *Pale skin.*
- *Restlessness.*
- *Rapid and thready pulse.*
- *Elevated breathing rate.*
- *Nausea.*
- *Bluish skin around the mouth.*

30. How do you treat for shock?
    - *Position the casualty.*
    - *Loosen tight-fitting clothing.*
    - *Keep the casualty from chilling (or overheating).*
    - *Keep the casualty calm.*

31. When should a patient not be placed in the standard shock position?
    When he or she has a head injury.

32. Identify some of the signs and symptoms of a head injury.
    - *Unequal pupils.*
    - *Slurred speech.*
    - *Confusion.*
    - *Sleepiness.*
    - *Loss of memory.*
    - *Loss of balance.*
    - *Headache.*
    - *Dizziness.*
    - *Vomiting.*
    - *Convulsions.*
    - *Paralysis.*
    - *Fluid draining from the ears, nose, or mouth.*

33. When should a patient not be put into a feet-elevated position?
    When the patient has a head injury.

34. Name three heat injuries.
    Heat exhaustion, heatstroke, and heat cramps.

35. What soldiers are most likely to suffer heat injuries?
    Soldiers not accustomed to heat; soldiers who are overweight; and soldiers who are already dehydrated because of diarrhea, use of alcohol, or insufficient intake of water.

36. Which of the heat injuries is the most severe?
    Heat stroke.

37. Specify the steps in dealing with heat stroke.
    - *Lower the body temperature as quickly as possible.*
    - *Carry the casualty to a cool, shaded area.*
    - *Immerse the casualty in cold water containing ice if possible or, if not, keep the entire body wet and fan casualty to lower temperature. Get medical aid as soon as possible.*

38. Identify the symptoms of heat exhaustion.
    - *Headache.*
    - *Excessive sweating.*
    - *Weakness.*
    - *Dizziness.*
    - *Pale and clammy skin.*
    - *Muscle cramps.*

39. How should you treat heat exhaustion?
    - *Lay the person in a shaded or cool area.*
    - *Loosen or remove outer clothing.*
    - *Elevate the feet.*
    - *Give cool water to drink if the person is conscious.*

40. What does the keyword COLD mean?
    C—Keep it clean.
    O—Avoid overheating.
    L—Wear loose clothing in layers.
    D—Keep it dry.

41. For what do the letters A-I-D-S stand?
    Acquired immunodeficiency syndrome.

42. Define AIDS.
    AIDS is the end-stage disease of the HIV (human immunodeficiency virus) infection, wherein the virus has attacked and weakened a person's immune system.

43. State the two most common ways HIV is transmitted.
    Through the sharing of intravenous drug needles and unprotected sexual contact.

44. AIDS is contagious. Can it be spread through casual contact?
    No.

45. If you have a soldier who overdoses on drugs, what is it important
    that you do?
    Get medical help immediately.

46. What are the signs of a fracture?
    • *Point tenderness.*
    • *Inability to move or sharp pain on movement.*
    • *Deformity.*
    • *Swelling.*
    • *Discoloration.*
    • *Unusual body position.*

47. State the two types of fractures.
    Open and closed (simple and compound).

48. What is a closed fracture?
    A break in the bone without a break in the skin covering the bone.

49. List several items of personal military equipment that you might
    use to make a splint.
    • *Rifle.*
    • *Bayonet.*
    • *Intrenching tool.*
    • *Tent poles.*
    • *Tent pegs.*
    • *Web belt.*

50. Describe the quickest way to splint a broken leg.
    Tie the broken leg securely to the unbroken leg.

51. Why is a fracture immobilized?
    *To prevent further damage (sharp bone fragments can cut tissue,
    blood vessels, and nerves).*

52. What should you remember when applying a splint?
    • *Stop the bleeding.*
    • *Splint it where it lies.*
    • *Immobilize the joints above and below the fracture.*

- *Use padding between splint and casualty.*
- *Check for circulation after making a tie.*
- *Use a sling if possible.*

53. What is the proven principle in the rules for splinting?
"Splint them where they lie."

54. When carrying an injured soldier for a long distance, which one-person carry would be best?
Pistol-belt carry.

55. List the four types of burns.
(1) Thermal.
(2) Electrical.
(3) Chemical.
(4) Laser.

56. Describe the three burn classifications.
First degree—reddening of the skin; second degree—blistering; and third degree—charred flesh.

57. State the two primary objectives when giving first aid to a burn casualty.
Prevent or lessen shock and prevent infection.

58. Describe the care of a serious burn.
Keep the burn clean, covered, and dry; seek medical aid.

59. Should serious burns be covered with an ointment or grease?
No, only with a sterile dressing.

60. Should large amounts of water be given to a person with a serious burn?
No.

61. Describe the first-aid procedures for a white phosphorus casualty.
Smother the flame by submerging the affected area in water, or pack with mud; then remove the particles by brushing or picking them out.

62. Name the signs of mild nerve agent poisoning.
    - *Unexplained runny nose.*
    - *Sudden headache.*
    - *Drooling.*
    - *Impaired vision.*
    - *Tightness in the chest or difficulty in breathing.*
    - *Localized sweating or twitching.*
    - *Stomach cramps.*
    - *Nausea.*

63. What are the symptoms of severe nerve agent poisoning?
    - *Strange or confused behavior.*
    - *Coughing and difficulty in breathing.*
    - *Pinpointed pupils.*
    - *Red tearing eyes.*
    - *Vomiting.*
    - *Severe muscular twitching.*
    - *Loss of bowel and bladder control.*
    - *Convulsions.*
    - *Unconsciousness.*
    - *Stoppage of breathing.*

64. Specify the first-aid treatment for a toxic chemical blood agent.
    There is no first-aid treatment.

65. Name the signs of carbon monoxide poisoning.
    - *Headache.*
    - *Dizziness.*
    - *Yawning.*
    - *Nausea.*
    - *Ringing in the ears.*
    - *Heart flutters.*

66. Specify the major cause of tooth decay and gum disease.
    Dental plaque.

67. Define "hypothermia."
    The lowering of body temperature due to loss of heat at a rate faster than it can be produced.

68. What foot injuries can cold and wet weather cause?
    Trench foot, immersion foot, and frostbite.

69. How can you prevent athlete's foot?
    Keep your feet clean, use foot powder, and change socks daily.

70. What are the three medical evacuation categories?
    (1) Urgent (within two hours).
    (2) Priority (within four hours).
    (3) Routine (within twenty-four hours).

71. What is a combat lifesaver?
    The combat lifesaver is a member of a nonmedical unit selected by the unit commander for additional training beyond basic first aid measures.

72. Does the combat lifesaver replace the medic?
    No. He complements the medic with training that bridges the gap between buddy aid and the medical specialist.

73. Describe the emergency bandage.
    It is an elastic bandage with a pad and a pressure bar.

74. How does the emergency bandage differ from a field dressing?
    The field dressing has a pad with two cloth tails. The emergency bandage has a pad with a pressure bar and an elastic wrap.

75. What is a CAT?
    Combat Application Tourniquet.

76. What are some common causes of shock?
    - *Dehydration.*
    - *Allergic reaction.*
    - *Loss of blood.*
    - *Reaction to the sight of a traumatic scene.*
    - *Traumatic injuries.*

77. What does the improved first aid kit (IFAK) contain?
    - *Exam gloves.*
    - *Trauma dressing.*
    - *Combat Application Tourniquet (CAT).*

- *Two-inch tape.*
- *Nasopharyngeal Airway (NPA).*
- *Kerlix dressing.*

78. Good personal hygiene and cleanliness do what?
    - *Protect against disease-causing germs.*
    - *Keep disease-causing germs from spreading.*
    - *Promote good health among soldiers.*
    - *Improve morale.*

79. What are the rules for avoiding illness in the field?
    - *Never consume food or drink from unauthorized sources.*
    - *Proper disposal of urine and feces.*
    - *Keep your fingers out of your mouth and nose.*
    - *Wash hands often.*
    - *Wash mess gear after each meal.*
    - *Clean your mouth and teeth at least daily.*
    - *Avoid insect bites.*
    - *Avoid sharing personal items with other soldiers.*
    - *Avoid leaving food scraps around.*
    - *Sleep when possible.*
    - *Exercise regularly.*

80. On the battlefield, casualties will fall into what three categories?
    - *(1) Casualties who will die regardless of receiving medical care.*
    - *(2) Casualties who will live regardless of receiving medical care.*
    - *(3) Casualties who will die if they do not receive timely and appropriate medical care.*

Flags

REFERENCES: AR 600-25, Salutes, Honors, and Visits of Courtesy;
AR 840-10, Heraldic Activities; FM 3-21.5, Drill and Ceremonies

1. Note three types of American flags, their size, and when they are
   flown.
   (1) The storm flag, 9 ½ by 5 feet, flown in inclement weather.
   (2) The post flag, 8 feet 11 ⅜ inches by 17 feet, flown on normal
   duty days.
   (3) The garrison flag, 20 by 38 feet, flown on holidays and impor-
   tant occasions.

2. What is a U.S. color?
   The U.S. flag trimmed on three sides with a golden yellow fringe
   2 ½ inches wide. It is primarily for indoor display.

3. List five places the American flag is flown twenty-four hours a
   day by specific legal authority as of January 1966.
   (1) U.S. Capitol in Washington, D.C.
   (2) Fort McHenry National Monument, Flag House Square in
   Baltimore, Maryland.
   (3) Francis Scott Key's grave.
   (4) The World War I Memorial in Worcester, Massachusetts.
   (5) The USS Arizona memorial in Pearl Harbor, Hawaii.

4. How is the flag honored if flown at night?
   It is properly illuminated.

5. For what do the colors red, white, and blue stand?
   Red for hardiness and valor; white for purity and innocence; and
   blue for vigilance, perseverance, and justice.

6. The American flag has how many stripes? How many of each
   color are there?
   Thirteen stripes; seven red and six white.

7. How is the American flag raised to the half-staff position and
   lowered at the end of the day?

*It is first hoisted to the top of the staff for an instant and then
lowered to the half-staff position. At the end of the day, it is again
hoisted to the top of the staff before being lowered for the day.*

8. When can the American flag be flown upside down?
   During an emergency and to alert someone.

9. Does the American flag ever salute (dip) when passing in review?
   No.

10. On what occasion, if any, would the national colors be dipped in
    salute or compliment?
    *Never on land, but they can be dipped at sea when two friendly
    ships of war meet.*

11. On post, where is building number one located?
    At the base of the flagpole.

12. What is a union jack?
    *A blue flag with fifty stars, used by naval vessels that are at
    anchor or tied to a pier. It represents the U.S. flag.*

13. When folded, what shape is the flag of the United States?
    A cocked hat.

14. What are the only two ways the flag can be displayed?
    Flat or hanging free.

15. How is the flag flown on Memorial Day?
    *It is flown at half-mast until noon, and at full mast from noon to
    retreat.*

16. Name the three places that the flag always flies at half-mast.
    *The Tomb of the Unknown Soldier, Arlington National Cemetery,
    and the Arizona Memorial at Pearl Harbor.*

17. When a president or past president dies, how long is the flag flown
    at half-mast?
    Thirty days.

18. What is the height of a military flagpole?
    50 feet, 60 feet (normal), or 75 feet.

19. There are four authorized decorative ornaments on the top of a flagstaff. Identify them.
    (1) Eagle (presidential).
    (2) Spearhead (Army flags).
    (3) Acorn (markers and pennants).
    (4) Ball (advertising and recruiting).

20. What is the ball on top of the flagpole called?
    The truck.

21. What does the "truck" on the flagpole represent?
    The shot heard around the world.

22. Define "hoist" as it pertains to flags.
    The height of a flag at its vertical edge, measured from top to bottom.

23. Define "fly" as it pertains to flags.
    The width of a flag at its horizontal edge, measured from left to right.

24. Describe the proper way to dispose of unserviceable flags.
    By burning or some other method that does not bring disrespect to the flag.

25. Name the types of military flags.
    - *Colors.*
    - *Standards.*
    - *Distinguished flags.*
    - *Ensigns.*
    - *Guidons.*
    - *Pennants.*

26. Describe the two types of U.S. colors used by the Army.
    The first has a 4-foot 4-inch hoist by 5-foot 6-inch fly and is displayed with the U.S. Army flag, positional colors, the Corps of Cadets colors, 1st Battalion 3rd Infantry colors, and the chapel flag. The second has a 3-foot hoist by 4-foot fly and is displayed

with the Army field flag, distinguishing flags, organizational
colors, institutional flags, and the chapel flag.

27. What streamer is in the preeminent position on the Army flag?
Yorktown 1781.

28. What is the first streamer on the Army flag?
Lexington 1775.

29. On what date was the Army flag dedicated?
14 June 1956.

30. Traditionally, who is responsible for the safeguarding, care, and
display of the unit's colors?
The command sergeant major.

31. What is a guidon?
A flag that identifies a company, troop, or battery.

32. In what position is the unit guidon after a preparatory command is
given?
The raised vertical position.

33. Guidons are authorized for detachments and separate platoons
with an authorized strength of how many military personnel?
Thirty or more.

34. Define "pennant."
A triangular flag used for various military purposes.

35. Identify the five basic U.S. flags used by the Army.
(1) Garrison.
(2) Post.
(3) Field.
(4) Storm.
(5) Interment flags.

36. How is the flag draped over a casket?
The union is at the head and over the left shoulder.

37. Who provides the flag presented to the next of kin at a military
    funeral?
    The postmaster general.

38. When the U.S. flag is displayed among a number of flags in a row,
    at what position is it placed?
    In the center and at the highest point.

39. When is the garrison flag normally displayed?
    On holidays and other important occasions.

40. What is the honor point on the U.S. flag?
    The blue field with the white stars.

41. When is the U.S. flag required to be displayed or carried?
    When other flags are displayed or carried.

42. What is a tabard?
    *A tabard is a rayon banner cloth that is attached to the tubing of a
    herald trumpet.*

43. What are the four types of finials found on U.S. flagstaffs?
    (1) Eagle.
    (2) Spearhead.
    (3) Acorn.
    (4) Ball.

44. The eagle flagstaff finial is only used on what flagstaffs?
    Presidential flagstaffs.

Guard Duty

REFERENCES: FM 21-75, Combat Skills of the Soldier;
FM 22-6, Guard Duty

1. List your chain of command as a sentinel.
   - *Commander of the relief.*
   - *Sergeant of the guard.*
   - *Officer of the guard.*
   - *Field officer of the day.*
   - *Commanding officer for the guard mount.*

2. Who within the guards assigns posts to the guards?
   Commander of the relief.

3. What action does the commander of the relief take when a guard is posted and relieved while live ammunition is used at the guard post?
   Supervises the loading, unloading, and clearing of weapons.

4. Who is responsible for the instruction, discipline, and performance of the guards?
   Commander of the guards.

5. Identify the general responsibilities of the commander of the guard.
   The instruction, discipline, and performance of the guard.

6. Define "supernumerary."
   Extra member of the guard who is used when needed to replace a guard or to perform special duties as prescribed by local directives.

7. What are the three types of guards?
   Interior, exterior, and special.

8. What is the purpose of an interior guard?
   To preserve order, protect property, and enforce military regulations.

9. Describe the three general orders that interior guards are required to memorize, understand, and comply with.
   - *I will guard everything within the limits of my post and quit my post only when properly relieved.*
   - *I will obey my special orders and perform all my duties in a military manner.*
   - *I will report any violations of my special orders, emergencies, and anything not covered in my instructions to the commander of the relief.*

10. What type of guard is normally used to protect a unit in a tactical environment?
    Exterior guard.

11. Identify three examples of exterior guards.
    Lookouts, listening posts, and outposts.

12. Name three types of guard duty.
    Interior guard, listening post, and prisoner guard.

13. Cite the two different types of guard mountings.
    Formal and informal.

14. Name two types of orders a guard can receive.
    General orders and special orders.

15. What do special orders define?
    Exactly what a guard must do on a particular post.

16. What two main qualifications must a soldier have before being eligible to pull guard?
    Know his or her general orders and be qualified with his or her individual weapon.

17. Must guards be qualified with the weapon they are carrying?
    Yes.

18. At what position should you hold your rifle while challenging?
    At port arms.

19. When challenging, what is the position of your weapon?
    Port arms if armed with a rifle, and raised pistol if armed with a handgun.

20. How does a posted guard with a rifle salute?
    Halts; faces music, person, or colors; and presents arms.

21. Is a guard required to salute when on a post that requires challenging?
    No.

22. What is a parole word?
    A secret word known only to the guards, commanders of the guards, and persons authorized to inspect the guards.

23. A countersign consists of two words. State them.
Challenge and password.

24. What is the penalty, in time of war, for illegal disclosure of the parole word or countersign?
Death or other punishment directed by a court-martial.

25. What are the hours of challenging?
Darkness or periods of poor visibility.

26. What specifies the time for challenging?
The special orders.

27. What is the normal length of time for a tour of guard duty?
Twenty-four hours.

28. How long is a guard normally posted?
Two to four hours.

29. How much time normally lapses from the relief of a guard until he or she is reposted?
Four hours.

30. Normally, how many reliefs are there in a guard?
Three.

31. What are your responsibilities as guard?
Your post and all government property in view.

32. Why should you never remove your clothing or equipment while on guard duty?
An emergency could arise that would require the presence of all guard personnel at once.

33. What punishment can be administered to a guard who sleeps while at his post?
Court-martial.

Guerrilla and Psychological Warfare

REFERENCES: FM 3-05.30, Psychological Operations;
FM 3-24, Counterinsurgency

1. What is special warfare?
   All the military and paramilitary measures and activities related to unconventional warfare, counterinsurgency, and psychological warfare.

2. What is insurrection?
   Acts of revolt against civil or political authority or the established government.

3. Define "guerrilla warfare."
   Military and paramilitary operations conducted on enemy-held territory by irregular, predominantly local forces.

4. Name the three phases of guerrilla operations or insurgency.
   Initial or organization phase, training and operational phase, and all-out assault on government forces (conventional warfare).

5. On what are guerrilla tactics based?
   Surprise, mobility, and dispersion of forces.

6. What are the requirements for a successful guerrilla operation?
   - *Cause.*
   - *Civil support.*
   - *Unity of effort.*
   - *Outside assistance.*
   - *Favorable terrain.*
   - *Effective leadership.*
   - *Use of propaganda.*
   - *Intelligence effort.*
   - *Discipline.*

7. Define "psychological warfare."
   The planned use of propaganda and exploitation of other actions, attitudes, and behavior of the enemy, neutral, or friendly foreign groups in such a way as to support the accomplishment of national aims and objectives.

8. What is meant by "black," "white," and "gray" propaganda?
Black—identifies the source incorrectly; white—identifies the source correctly; gray—does not identify the source.

9. What are covert operations?
Those that do not disclose the source of origin.

10. Name the three types of media used in psychological operations.
Audio, visual, and printed material.

Hand Grenades

REFERENCES: FM 3-23.30, Grenades and Pyrotechnic Signals;
TRADOC PAM 600-4, IET Soldier's Handbook

1. Name the three main parts of a grenade.
   Body, filler, and fuse assembly.

2. What are the five uses of hand grenades?
   (1) Causing casualties.
   (2) Signaling.
   (3) Screening.
   (4) Incendiary effects.
   (5) Riot control.

3. Describe the color and marking on a fragmentation hand grenade.
   Olive drab with yellow marking around the grenade body.

4. What is the length of time delay in hand grenades?
   1.4 to 3 seconds for chemical-burning grenades; 4 to 5 seconds for casualty-producing grenades.

5. When you grip the hand grenade, where should the safety lever be?
   Under the thumb, between the first and second joint.

6. Describe the sequence of events after the safety lever on a grenade is released.
   • *Lever flies free of grenade.*
   • *Striker rotates and strikes primer.*
   • *Primer emits an intense flash of heat, which starts time delay element.*
   • *Delay element sets off detonator or igniter, setting off filler.*

7. Describe the filler of a hand grenade.
   Either a high explosive or a chemical; type of filler determines its use.

8. If you pull the pin on a grenade, is it all right to reinsert the pin if you are careful?
   No. Once you pull the pin on a grenade, the grenade must be thrown.

9. If you remove the safety clip from a grenade, is it all right to reattach the safety clip if you are careful?
   Yes, if the safety pin is still in place.

10. When inspecting a grenade, what are some of the things you should look for?
    • *Missing safety clip or pin.*
    • *Cracked body.*
    • *Broken fuse lug.*
    • *Bent or broken safety lever.*
    • *Cracked pull ring.*
    • *Loose fuse.*

11. What are the basic types of hand grenades used by the U.S. Army?
    • *Fragmentation.*
    • *Chemical.*
    • *Offensive.*
    • *Illumination.*
    • *Non-lethal.*
    • *Practice.*

12. Which U.S. hand grenade is the heaviest?
    The AN-M14 TH3 Incendiary.

13. What two fuse types are found in grenades?
    Detonating and ignition.

14. What type of fuse is used in a chemical grenade?
    Igniting.

15. How many grenades can the standard ammunition pouch carry?
    Five fragmentation grenades—two in the sleeves and three inside the pouch.

Health Promotion, Risk Reduction, and Suicide Prevention

REFERENCE: DA PAM 600-24

1. To reduce suicides, leaders should do what?

   Promote a climate of support, minimize stigma, and encourage help-seeking behavior.

   Understand leader responsibilities regarding suicide prevention, intervention, and postvention.

   Take a personal interest in subordinates' personal lives and provide support where needed.

   Teach suicide prevention to all.

   Implement the battle-buddy system and foster a sense of responsibility in solders to provide care and support to peers.

2. To reduce suicides, soldiers should do what?

   Care for your buddy.

   Seek out your buddy for advice, protection, and support.

   Recognize that seeking help is a sign of strength.

   Report any concerns that a buddy may harm themselves.

Leadership

REFERENCES: FM 6-22, Army Leadership; FM 7-22.7, The NCO Guide;
DA PAM 350-58, Leader Development For America's Army

1. What is the Army's first responsibility?
   To fight and win the nation's wars.

2. According to FM 6-22, what is the foundation of the Army?
   Confident and competent leaders of character.

3. What is a leader?
   Anyone who is responsible for supervising people or accomplishing a mission that involves other people.

4. Define leadership.
   The process of influencing others to accomplish the mission by providing purpose, direction, and motivation.

5. What is the purpose of leadership?
   To accomplish the mission and improve the organization.

6. What are the four indicators of good leadership?
   (1) Proficiency.
   (2) Discipline.
   (3) Unit cohesion.
   (4) High morale.

7. What are the nine leadership competencies?
   (1) Communications.
   (2) Supervision.
   (3) Teaching and counseling.
   (4) Soldier team development.
   (5) Technical and tactical proficiency.
   (6) Decisionmaking.
   (7) Planning.
   (8) Use of systems.
   (9) Professional ethics.

8. What is the Army leader's most important challenge?
   Leadership in combat.

9. What is the Warrior Ethos?
The professional attitude and beliefs characterizing the American soldier that is grounded on the refusal to accept failure.

10. Define a combat-effective, combat-ready unit.
A unit that will accomplish any assigned mission for which it has been organized, equipped, and trained to do; in the shortest possible time; with a minimum expenditure of means; and with the least amount of confusion.

11. Name the two most important responsibilities of a leader.
Mission accomplishment and welfare of soldiers.

12. Name the most important thing a leader must know or have.
Tactical skills.

13. What is the difference between technical and tactical knowledge?
Technical knowledge is the knowledge needed to operate a piece of equipment. Tactical knowledge is the knowledge needed to effectively employ men and equipment.

14. What are the two areas that make up tactical skills that a leader should know?
Doctrine and field craft.

15. What are the physical attributes of a leader?
Good health, physical fitness, and professional military bearing.

16. List the mental attributes of a leader.
- *Will.*
- *Self-discipline.*
- *Initiative.*
- *Judgment.*
- *Self-confidence.*
- *Intelligence.*
- *Cultural awareness.*

17. What are the emotional attributes of a leader?
Self-control, balance, and stability.

18. What do self-control, balance, and stability help you do?
Make the right ethical choices.

19. What are the four categories of leader skills?
    (1) Interpersonal.
    (2) Conceptual.
    (3) Technical.
    (4) Tactical.

20. What are a leader's conceptual skills?
    Critical reasoning, ethical reasoning, and creative thinking.

21. What is essential to mastering leader skills?
    An understanding of human beings.

22. What are the three levels of leadership?
    Direct, organizational, and strategic.

23. Leadership is developed largely as a result of what?
    Experience, environment, and training.

24. How are Army leaders trained?
    Institutional training, operational assignments, and self-development.

25. What items are entrusted to Army leaders?
    - *People.*
    - *Facilities.*
    - *Equipment.*
    - *Training.*
    - *Resources.*

26. State the leadership fundamentals.
    - *Provide influence.*
    - *Provide purpose.*
    - *Provide direction.*
    - *Provide motivation.*
    - *Operate.*
    - *Improve.*

27. What are the three categories of leader actions?
    Influencing, operating, and improving.

28. According to FM 6-22, you cannot be an effective leader until you do what?
Apply what you know; act; and do what you must.

29. Describe leadership actions and orders.
Anything a leader does or says to influence and direct his or her command.

30. What does a good leader do once he or she has issued an order or orders?
Check to see that the order or orders are being properly carried out.

31. Mention six of the eleven principles of leadership.
(1) Know yourself and seek self-improvement.
(2) Be technically and tactically proficient.
(3) Seek responsibility and take responsibility for your actions.
(4) Make sound and timely decisions.
(5) Set the example.
(6) Know your soldiers and look out for their well-being.
(7) Keep your soldiers informed.
(8) Develop a sense of responsibility in your subordinates.
(9) Ensure that the task is understood, supervised, and accomplished.
(10) Train your soldiers as a team.
(11) Employ your unit in accordance with its capabilities.

32. What does the term "power down without powering off" mean?
It means you should give subordinate leaders the authority to do assigned tasks and check on them without getting in their way.

33. Name the five types of leadership styles.
(1) Directing.
(2) Participating.
(3) Delegating.
(4) Transformational.
(5) Transactional.

34. What is the problem with the transactional style of leadership?
It gets only a short-term commitment from subordinates and discourages risk-taking and innovation.

35. Describe the transformational leadership style.
   Transforming subordinates by challenging them to rise above their immediate needs and self-interest.

36. Name the seven Army values. (L-D-R-S-H-I-P)
   (1) Loyalty.
   (2) Duty.
   (3) Respect.
   (4) Selfless service.
   (5) Honor.
   (6) Integrity.
   (7) Personal courage.

37. What are the two other values sometimes added to the list of Army values?
   Teamwork and discipline.

38. What is courage?
   The ability to act as you should in spite of fear.

39. What are the two components of personal courage?
   Physical and moral courage.

40. What are the three stages of team building?
   Formation, enrichment, and sustainment.

41. Describe some leader actions that are essential to team building.
   - *Setting the example.*
   - *Assisting soldiers with problems.*
   - *Establishing the buddy system.*
   - *Providing a stable unit climate.*
   - *Emphasizing safety awareness.*
   - *Devising more challenging training.*
   - *Training as a unit.*
   - *Train for combat (train as you will fight).*
   - *Developing soldier and unit goals.*
   - *Establishing clear lines of authority.*
   - *Building pride and accomplishment.*
   - *Self-evaluations.*

42. Name some conditions that will undermine teamwork.
    • *Continuous operations.*
    • *Enemy operations.*
    • *Casualties.*
    • *Boredom.*
    • *Rumors.*

43. What is discipline?
    A moral, mental, and physical state in which all ranks respond to the will of the leader whether he is there or not.

44. What is considered the core of a soldier?
    Moral discipline.

45. Define ethics.
    Rules or standards that guide individuals or groups to do the moral or right thing.

46. Identify the cornerstone of the professional Army ethic.
    Character and adherence to the principles the Army values embody.

47. Our military standard of ethics comes from what documents?
    • *Oath of Allegiance to the Constitution.*
    • *Code of Conduct.*
    • *Uniform Code of Military Justice (UCMJ).*
    • *Code of Ethics for Government Service.*

48. Character is made up of what two interactive parts?
    Values and attributes.

49. What two things are the basis or character of a leader?
    Army values and leader attributes.

50. How do you demonstrate your character?
    Through your behavior.

51. What two traits define an Army leader?
    Character and competence.

52. Leaders develop character in others by doing what?
    * *Teaching Army values and demonstrating attributes.*
    * *Reinforcing Army values and leader attributes.*
    * *Shaping the organization's ethical climate.*

53. What are standards?
    Principles or rules by which behavior and task completion are measured against successful accomplishment.

54. Explain proficiency (competence).
    The ability of a soldier or unit to accomplish missions to the prescribed standard.

55. What three things fall under a leader's interpersonal skills?
    Communicating, supervising, and counseling.

56. What is supervision?
    Keeping a grasp on a situation and ensuring that plans and policies are being followed; the art of checking without undue harassment.

57. What is authority?
    Authority is the legitimate power of leaders to direct subordinates or to take action within the scope of their positions.

58. May an enlisted soldier be a commander?
    Yes.

59. May a civilian exercise command?
    No.

60. As an NCO, you have two types of authority. What are they and from where do they come?
    (1) Command authority, which comes from the rank and position you hold; it is delegated down to you through the chain of command.
    (2) General military authority, which is the authority to act in the absence of other leaders. It stems from the Constitution, Army regulations, the UCMJ, and other written policies and letters of instruction.

61. NCOs have duties and responsibilities. Contrast the difference between a duty and a responsibility.
Responsibility is being accountable for what you do or do not do. A duty stems from a responsibility; it is something you must do because of the rank or position you hold.

62. What are the three types of duties that NCOs have?
Specified, directed, and implied.

63. When arriving at a new unit, what questions should you find answers to?
* *What is the unit's mission?*
* *How does this mission fit in with the mission of the next higher organization?*
* *What are the standards the unit must meet?*
* *What resources are available?*
* *What is the current state of morale?*
* *Who reports directly to me?*
* *What are the strengths and weaknesses of key subordinates?*
* *Who are the key people outside the unit that support the unit's mission?*

64. How do NCOs gain the respect and confidence of soldiers?
By demonstrating technical and tactical proficiency and by caring for soldiers and their families.

65. What is the NCO Vision?
An NCO Corps, grounded in heritage, values, and traditions, that embodies the warrior ethos, values perpetual learning, and is capable of leading, training, and motivating soldiers.

66. Name some basic responsibilities of an NCO.
* *Training soldiers.*
* *Maintaining discipline.*
* *Maintaining government property.*
* *Ensuring the welfare of soldiers.*
* *Executing the mission.*
* *Supporting the chain of command.*

67. What is the NCO's principal duty and responsibility?
Training.

68. What are the team-building stages?
Formation stage, enrichment stage, and sustainment stage.

69. What is sergeant's training time?
It is hands-on, practical training given by NCOs. It should be dedicated time given to essential war-fighting tasks.

70. The *NCO Guide* lists four ways to become a better leader. Name them.
(1) Know your job.
(2) Know yourself.
(3) Know your soldiers.
(4) Be honest.

71. State the first duty of a soldier in regards to orders and instructions.
Obedience.

72. Name the four soldierly values.
Courage, candor, competence, and commitment.

73. List the main parts of a five-paragraph field order.
(1) Mission.
(2) Situation.
(3) Execution.
(4) Service support.
(5) Command and signal.

74. What is communication?
It is an "influencing action" and an "interpersonal skill" involving oral, written, and listening means to exchange ideas or information so that they are clearly understood.

75. What are the two key parts to communication?
Sending and receiving.

76. What are the four communications categories?
Speaking, reading, writing, and listening.

77. What is the key difference between one-way and two-way communication?
One-way communication may not provide a complete picture.

78. What term defines those things that interfere with communication?
Barriers.

79. What is tact?
Speaking and acting in an inoffensive manner in order to maintain good relationships with others.

80. What is meant by management?
The process of using people, money, material, time, and facilities in the most economical way to accomplish the mission.

81. List the steps in the problem-solving process.
(1) Identify the problem.
(2) Identify facts and assumptions.
(3) Generate alternatives.
(4) Analyze the alternatives.
(5) Compare the alternatives.
(6) Make and execute your decision.
(7) Assess the results.

82. What action is most likely to increase the chance of a desired behavior?
Positive reinforcement.

83. Name some negative tools of motivation.
• *Mass punishments.*
• *Threats of punishment.*
• *Verbal and written reprimands.*
• *Action under the UCMJ.*

84. List some positive tools of motivation.
• *Establishing clear goals and standards.*
• *Setting the example.*
• *Making tasks challenging.*
• *Making unit and soldier needs coincide if possible.*
• *Promotions.*
• *Awards and rewards.*

85. What is counseling?
Counseling is subordinate-centered communication that produces a plan outlining actions necessary for subordinates to achieve individual or organizational goals.

86. Are leaders required to counsel subordinates?
    Yes.

87. When should a leader counsel a subordinate?
    As the situation requires or at least quarterly.

88. State some basic counseling skills.
    Listening, watching, responding, questioning, and guiding.

89. Cite the key to holding a successful counseling session.
    Being prepared.

90. What is the most crucial phase of a counseling session?
    The opening or first few minutes.

91. What are the two primary counseling areas outlined in AR 22-100?
    (1) Event-oriented.
    (2) Performance and professional growth.

92. What areas of counseling come under performance and professional growth counseling?
    Personal performance evaluations (OERs and NCOERs) and planning for professional goals, such as schooling and assignments.

93. What types of counseling events are encountered in event-oriented counseling?
    - *Specific instance.*
    - *Event oriented.*
    - *Reception and integration.*
    - *Crisis.*
    - *Referral.*
    - *Promotion.*
    - *Adverse separation.*
    - *Performance and professional growth.*

94. Specific incident counseling addresses what areas?
    Individual performance.

95. If you talk to one of your soldiers about a test score, what type of counseling are you conducting?
    Performance.

96. If you talk to a soldier about a family problem, what type of counseling are you conducting?
    Personal.

97. What are the three basic approaches to counseling?
    Directive, nondirective, and combined.

98. Name the two means of conducting counseling.
    Formal and informal.

99. State the four parts of a counseling session.
    (1) Identifying the need.
    (2) Planning and preparation.
    (3) Conducting the session.
    (4) Following up.

100. What is a DA Form 4856-E used for?
    It is the standard counseling form used for development, event, performance, and professional growth counseling.

101. During the discussion phase of a counseling session, what is the most important role of the counselor?
    To guide the session.

102. State the four common counseling errors.
    (1) Talking too much.
    (2) Giving unnecessary or inappropriate advice.
    (3) Not listening.
    (4) Projecting personal likes, dislikes, and prejudices.

103. What are the barriers to listening?
    • *Thinking of a response while the other person is still talking.*
    • *Distractions such as anger.*
    • *Thinking of other subjects or things during the conversation.*
    • *Giving in to personal or emotional thoughts about the speaker.*

104. List the four counseling techniques.
    Suggesting, recommending, persuading, and advising.

105. List the five "W" questions that fit many counseling situations.
    Who, what, when, where, and why.

106. Should a leader try to solve every problem?
No. He or she should encourage the counseled person to solve his or her own problem and, if appropriate, advise the person on how to do so.

107. What are the five main silent cries for help that a leader should recognize and answer?
(1) A good performer begins to constantly perform below par.
(2) A normally attentive soldier suddenly displays a lack of attentiveness or concentration.
(3) A moderate drinker begins to drink excessively.
(4) A good soldier refuses to follow instructions or becomes involved in deliberate acts of misconduct.
(5) A soldier lingers after meetings or formation to talk, asking such questions as, "What if a person has a problem . . . ?"

108. Human needs are important in understanding soldiers. Cite the four basic human needs.
Physical, security, social, and higher needs.

109. What are the areas defined as higher needs?
Religion, increased competence, serving a worthwhile cause, feeling needed.

110. Can a soldier be motivated by higher needs before his or her other needs are met?
Yes.

111. To develop an effective counseling style, you should develop what five characteristics?
Purpose, flexibility, respect, communication, and support.

112. What qualities are needed for leaders to be effective counselors?
• *Respect for subordinates.*
• *Self and cultural awareness.*
• *Empathy.*
• *Credibility.*

113. Describe esprit de corps.
The spirit of a unit.

114. Mention some activities that enhance unit pride and spirit in a unit.
   Military activities, sports activities, social activities, spiritual activities.

115. What are some ways to control rumors?
   • *Stress honesty.*
   • *Inform soldiers.*
   • *Maintain open lines of communication.*
   • *Identify and counsel those who spread rumors.*

116. Who is the key to the successful orientation of a new soldier?
   The squad leader.

117. Define morale.
   The individual's state of mind; how the individual feels toward his or her unit and leaders; and the individual's confidence in his or her own ability.

118. Name five things in a company that can greatly affect the morale of the unit.
   • *Dining facility.*
   • *Military justice.*
   • *Mail.*
   • *Living conditions.*
   • *Initial welcome to the unit.*
   • *Leadership.*
   • *Training.*
   • *Missions.*
   • *Social activities.*

119. Must a commander have an open-door policy?
   Yes.

120. Name some of the stresses that can cause ineffectiveness in a unit, especially in a combat situation.
   Rumors, fear, hunger, illness, anxiety, fatigue, and enemy fire.

121. Name two elements that must be present before a search of a soldier's area can be considered lawful.
   Authorization and probable cause.

122. Which agency on post is most appropriate for contacting relatives in an emergency?
The American Red Cross.

123. What term describes making judgments about individuals on the basis of presumed physical or group characteristics?
Stereotyping.

124. Identify the first step a leader should take to prevent racially motivated incidents.
Recognize incidents of unrest.

125. List four types of pressure a soldier must resist in order to avoid becoming unethical.
Self-interest, peer, subordinate, and senior.

126. One night while checking the billets area of your platoon you observe a soldier smoking a substance you believe to be marijuana. What should you do?
Collect the substance and notify the unit commander immediately.

127. If a soldier is determined to be a rehab failure, he or she can be processed for separation under what authority?
AR 635-200, Enlisted Personnel (Personnel Separations).

128. What are the steps involved in developing a leadership training program or improving a unit?
(1) Assess the situation and gather information.
(2) Analyze information to identify problems.
(3) Develop plan of action.
(4) Execute the plan of action.
(5) Evaluate progress.

129. What is the primary reason for conducting an inspection?
To evaluate the readiness of equipment and personnel.

130. Cite the most important aspect of an inspection.
How it is conducted.

131. What is the Army's management philosophy?
    - *Do the right things.*
    - *The right way.*
    - *For the right reasons.*
    - *And constantly strive for improvement.*

132. What does the Army expect its leaders to be able to do?
    - *Anticipate, manage, and exploit changes.*
    - *Be versatile enough to operate successfully in war and operations other than war.*
    - *Exemplify the highest professional and ethical standards.*
    - *Uphold the dignity of each individual.*
    - *Display technical and tactical proficiency.*
    - *Possess teaching, coaching, and counseling skills.*
    - *Build cohesive teams.*
    - *Communicate effectively while stimulating confidence, enthusiasm, and trust.*
    - *Accurately assess situations, solve problems, and act decisively under pressure.*
    - *Show initiative, plan thoughtfully, and take reasoned, measured risks to exploit opportunities.*
    - *Clearly provide purpose, direction, motivation, and vision while executing operations.*

133. What are the Army Values?
    Loyalty, Duty, Respect, Selfless Service, Honor, Integrity, Personal Courage, and Discipline.

134. What are the five dimensions of strength?
    Physical, emotional, social, family, and spiritual.

M16 Series Rifle and Rifle Marksmanship

REFERENCES: FM 3-22.9, Rifle Marksmanship;
TRADOC PAM 600-4, IET Soldier's Handbook

1. Describe the features of the M16 rifle.
   - *5.56mm.*
   - *Magazine-fed.*
   - *Gas-operated.*
   - *Air-cooled.*
   - *Semiautomatic or automatic.*
   - *Handheld.*
   - *Shoulder-fired.*

2. What eight types of ammunition can be used with the M16?
   (1) Ball, M193.
   (2) Tracer, M196.
   (3) Dummy, M199.
   (4) Blank, M200.
   (5) Ball, M855.
   (6) Tracer, M856.
   (7) SRTA, M862.
   (8) Armor Piercing, M995.

3. The M16 has two adjustable sights. Which is used to make elevation adjustments?
   The front sight.

4. Each click of the adjustable front sight equals how many centimeters?
   2.8 centimeters per every 100 meters of range (same for each notch of windage on the rear sight).

5. What are some differences between the M16A1 and the M16A2?
   - *Battlefield zero is 250 meters for the M16A1 and 300 meters for the A2.*
   - *The A2 has a bust position for the fire selection lever; the A1 does not.*
   - *On the A1, the long-range aperture is marked with an L; on the A2, it is not.*

6. The rear sight of the M16A1 has two apertures. When do you use the one marked with an "L?"
   When zeroing the weapon and when firing at targets beyond 350 meters.

7. What is battle sight zero?
   Sight settings that will cause a hit at the point of aim for a given range; the range that a round reenters the line of sight—250 meters for the M16A1 and 300 meters for the M16A2.

8. What is the weight of the M16A1?
   6.5 pounds; 7.5 pounds with loaded twenty-round magazine.

9. State the maximum effective range of the M16A1.
   460 meters.

10. Define maximum effective range.
    The greatest distance at which the weapon may be expected to inflict casualties.

11. At what range is the M16A1 zeroed?
    At a 25-meter range.

12. Name the eight steps in the functioning cycle of the M16.
    (1) Firing.
    (2) Unlocking.
    (3) Extracting.
    (4) Ejecting.
    (5) Cocking.
    (6) Feeding.
    (7) Chambering.
    (8) Locking.

13. Define stoppage.
    Any unintentional interruption in the cycle of function.

14. What is immediate action?
    The application of probable remedy to reduce a stoppage without investigation of the cause.

15. Explain a malfunction.
The weapon ceasing to fire because of a stoppage resulting from a mechanical failure of the weapon, magazine, or ammunition.

16. Identify the three major categories of malfunctions.
Failure to feed, chamber, or lock; failure to fire the cartridge; and failure to extract and eject.

17. Identify the positions for the selector lever on the M16.
Automatic, semiautomatic, and safe.

18. When clearing the M16, name the first two steps you should take.
Place the selector switch on safe; remove the magazine.

19. Identify the fundamentals for good marksmanship.
Steady position, proper sight picture, breath control, trigger control.

20. For what does the key word SPORTS stand?
   • *Slap.*
   • *Pull.*
   • *Observe.*
   • *Release.*
   • *Tap.*
   • *Shoot.*

21. What is the weight of the M16A2 without sling or magazine?
7.7 pounds.

22. What is the length of the M16A2?
Rifle length with compensator is 39.63 inches; rifle with bayonet is 44.84 inches.

23. Explain the purpose of the compensator.
It helps keep the muzzle down during firing, and it breaks up the muzzle flash.

24. What is the recommended basic load, with thirty-round magazines, for the M16A2?
210 rounds.

25. Name the type of ammunition that you should not attempt to fire in the M16A2.
    Corroded ammo, dented cartridges, cartridges with loose bullets, cartridges exposed to extreme heat, and cartridges with pushed-in bullets (short rounds).

26. Specify the muzzle velocity of ball ammunition fired from the M16A2.
    3,100 feet per second.

27. What are the maximum effective rates of fire for the M16A2?
    Sustained—twelve to fifteen rounds per minute; semiautomatic—forty-five rounds per minute; three-round burst—ninety rounds per minute.

28. On the 25-meter range, a one-notch movement of the front sight will move the bullet how far?
    .375 inch, up or down.

29. On the 25-meter range, a one-notch movement of the rear sight will move the bullet how far?
    .125 inch, right or left.

30. Which rear aperture is used when zeroing the M16A2?
    The unmarked aperture.

31. How do you perform maintenance on the M16A2 magazine?
    Wipe dirt from the tube, spring, and follower and lubricate only the spring.

32. List the three basic steps to proper rifle maintenance.
    Inspect, clean, and lubricate.

33. For what two things do you inspect your weapon?
    Cleanliness and serviceability.

34. What is used to clean the M16A2?
    CLP, swabs, pipe cleaners, brushes, patches, and cleaning rods.

35. Mention the three levels of lubrication for the M16A2.
    Light lube, general lube, and one-drop lube.

36. What is the maximum effective range of the M16A2?
Point targets—550 meters; area targets—800 meters.

37. State the maximum range of the M16A2.
3600 meters.

38. What is the M4 carbine?
The M4 carbine is a shortened variant of the M16. It has a collapsible stock and a shorter barrel.

39. What is the M68 sight?
The M68 sight is a reflex, nontelescopic sight using a red dot. It is designed for two-eyes open sighting and for use with the M16 and M4 carbine.

40. What are the advanced firing positions for the M16 series rifle?
   • *Alternate prone firing position.*
   • *Kneeling supported firing position.*
   • *Kneeling unsupported firing position.*
   • *Standing firing position.*
   • *Modified supported firing position.*
   • *Urban Operations firing positions.*
   • *Modified automatic and burst fire positions.*

41. What are the four combat fire techniques?
(1) Rapid semiautomatic fire.
(2) Automatic or burst fire.
(3) Suppressive fire.
(4) Quick fire.

42. What is the weight of the M4 Carbine?
6.49 pounds.

43. What is the difference between the M16A2/A3 and the M16A4?
The addition of the M5 rail adapter and a detachable carrying handle.

44. Is the muzzle velocity of the M4 Carbine different than the M16A2/A3?
Yes. The muzzle velocity of the M4 Carbine is 2,970 feet per second and the velocity for the M11A2/A3 is 3,100 feet per second.

M18A1 Claymore Mine

REFERENCES: FM 3-22.23, Antipersonnel Mine M18 and M18A1

1. What is the effective frontal range of the M18A1?
   50 meters.

2. Describe the primary use of the M18A1 Claymore mine.
   It is designed for use against massed infantry attacks.

3. Specify the method of employment that can be used with the Claymore mine.
   Controlled (fired by the operator).

4. List the three ways to prime the M18A1.
   With an electric cap, with a nonelectric cap, or with detonating cord.

5. During the installation of the Claymore mine, who should keep the M57 firing device?
   The individual installing the mine.

6. What is the explosive charge in the Claymore mine?
   1.5 pounds of Composition 4 (C4).

7. When a Claymore mine explodes, what causes the injuries?
   Blast and seven hundred small steel balls.

8. Describe the fan pattern for a Claymore mine.
   The pattern is an arc of about 60 degrees, 2 meters high, and 50 meters long.

9. What is the effective forward range of the blast?
   The balls are effective out to about 100 meters and dangerous up to 250 meters.

10. What components make up the M7 bandoleer?
    One M18A1 mine, one M40 test set, one M57 firing device, one electrical blasting cap, 100 feet of firing wire, an instruction sheet, and insulation tape.

11. What is the weight of the M18A1 mine?
    3.5 pounds.

M60 Machine Gun

REFERENCES: FM 3-21.75, Combat Skills of the Soldier;
TRADOC PAM 600-4, IET Soldier's Handbook;
FM 3-22.67, Machine Gun 7.62mm, M60

1. Describe the M60 machine gun.
   A 7.62mm, air-cooled, belt-fed, gas-operated, automatic tripod- or bipod-fired weapon.

2. Identify the six major groups of the M60.
   (1) Stock group.
   (2) Operating group.
   (3) Buffer group.
   (4) Trigger housing group.
   (5) Barrel group.
   (6) Receiver group.

3. What is the length of the M60?
   43.5 inches.

4. Cite the weight of the M60.
   23 pounds.

5. Name the five types of ammunition used with the M60.
   (1) Ball.
   (2) Tracer.
   (3) Armor piercing.
   (4) Dummy.
   (5) Blank.

6. Indicate the basic load of ammunition for the crew of the M60.
   Six hundred to nine hundred rounds.

7. What feature of the M60 allows the operator to change barrels rapidly?
   A fixed headspace.

8. In what position is the bolt when you fire the M60?
   Open-bolt position (bolt lock to the rear).

9. Why is the M60 fired from the open-bolt position?
For better cooling; it allows airflow through the bore as well as around the external surfaces of the barrel.

10. What is the maximum rate of fire for the M60?
550 rounds per minute (approximately).

11. What is the sustained rate of fire for the M60?
One hundred rounds per minute.

12. How many rounds per minute are fired with the M60 during rapid fire?
Two hundred rounds per minute.

13. In what position is the bolt for loading and unloading the M60?
Open-bolt position.

14. Describe the primary purpose of a range card.
Permits the gunner to place fire on designated targets during periods of limited visibility.

15. What are the rates of fire for the M60?
Sustained rate is one hundred rounds per minute, rapid rate is two hundred rounds per minute, and cyclic rate is 550 rounds per minute.

16. When firing the M60, what is the searching technique?
Moving the muzzle of the weapon up and down to distribute fire in depth.

17. What is grazing fire?
Horizontal knee-to-waist-high fire on the enemy.

18. Specify the maximum range of grazing fire with the M60.
600 meters.

19. What is the maximum effective range of the M60?
1,100 meters.

20. Indicate the maximum range of the M60.
3,725 meters.

21. What is the muzzle velocity of the M60?
    2,800 feet per second.

22. What should be done if a round gets stuck in the barrel during combat operations?
    The crew should change the barrel, reload, and continue firing.

23. Identify two common malfunctions of the M60.
    Sluggish operation and runaway gun.

24. Indicate the first action you take with the M60 if you have a runaway gun.
    Break the ammo belt.

25. What is the tracer burnout range of the M60?
    900 meters.

26. How many barrels are issued with the M60?
    Two.

27. What form is used to report faults on the M60?
    DA Form 2404.

M72A2 Light Antitank Weapon (LAW)
M202 Rocket Launcher

REFERENCES: FM 3-21.75, Combat Skills of the Soldier;
FM 3-22.25, Light Antiarmor Weapons; TC 23-2,
66mm Rocket Launcher; TM 3-23.25, Shoulder-Launched Munitions

1. Describe the M72 (LAW).
   The M72 is a shoulder-fired, single-shot, short-range antitank weapon.

2. What is the maximum range of the M72 (LAW)?
   1,000 meters.

3. Specify the maximum effective ranges of the M72A2 or M72A3 LAW.
   Stationary targets—200 meters; moving targets—165 meters.

4. What is the minimum arming range of the M72 (LAW)?
   10 meters.

5. To engage a target with the M72A2 LAW, what must you first determine?
   The range.

6. Before firing the M72A2 LAW, for what must you check?
   • *Check all seals.*
   • *Check for damage to the tube.*
   • *Ensure that the rubber detent and trigger boot are not torn.*
   • *Check for the words "with coupler" on the data plate.*

7. What caliber is the M72A2 LAW?
   66mm.

8. What is the difference between the M72 (LAW) and the M202 (Flash) rocket launcher?
   • *The M72 is a single-shot, throwaway launcher. The M202 is reloadable.*
   • *The M72 fires a round with a high-explosive charge. The M202 fires a round with an incendiary charge.*

9. What is the effective range of the M202A1?
   Area targets are 750 meters, and point targets are 200 meters.

10. What is the bursting radius of the M202A1 warhead?
    20 meters.

M136, AT4 Antitank Rocket

REFERENCE: FM 3-22.25, Light Antiarmor Weapons

1. Describe the AT4.
   The AT4 is a 84mm, light-weight, single-shot, man-portable, self-contained, shoulder-fired antitank rocket.

2. What is the minimum engagement range for the AT4?
   30 meters.

3. What is the minimum effective range of the AT4?
   300 meters.

4. What is the maximum range of the AT4?
   2,100 meters.

5. What FM covers the AT4?
   FM 3-23.25.

6. What is the overall weight of the AT4?
   15 pounds.

7. What is the danger area behind the AT4?
   5 meters in combat and 60 meters in training.

M141 Bunker Defeat Munition (BDM)

REFERENCES: FM 3-21.75, The Warrior Ethos and Soldier Combat Skills;
FM 3-23.25, Shoulder Launched Munitions

1. Describe the M141 Bunker Defeat Munition.
   *It is a lightweight, self-contained, man-portable, high-explosive,
   disposable, shoulder-launched, multipurpose, assault weapon.*

2. What is the M141 BDM?
   It is an 83mm, high-explosive, dual-mode, assault rocket round.

3. Must the M141 BDM be extended before firing?
   Yes.

4. What is the weight of the M141 BDM?
   15.7 pounds.

5. What is the maximum effective range of the M141 BDM?
   300 meters.

6. What is the minimum arming range of the M141 BDM?
   15 meters.

7. The M141 BDM is a shoulder-fired weapon. Can it be fired from
   the left shoulder?
   No, it only fires from the right shoulder.

Javelin, Medium Antiarmor Weapon System

REFERENCES: FM 3-21.75, The Warrior Ethos and Soldier Combat Skills;
FM 3-22.37, Javelin, Medium Antiarmor Weapon System

1. What is the Javelin?
   The Javelin is a fire and forget, crew-served, 126mm antitank missile.

2. What are the two components that make up the Javelin?
   (1) A reusable command launch unit (CLU).
   (2) A missile in a disposable launch tube assembly.

3. Does the Javelin command launch unit (CLU) have a secondary use?
   Yes, as a surveillance and reconnaissance tool.

4. What is the weight of the Javelin?
   49.5 pounds.

5. What is the maximum effective range of the Javelin?
   2,000 meters.

6. What is the minimum effective range of the Javelin?
   65 meters.

M203 and M320 Grenade Launcher

REFERENCES: FM 3-21.75, Combat Skills of the Soldier;
TM 3-22.31, 40mm Grenade Launcher, M203 and M79

1. Describe the M203 grenade launcher.
   A 40mm, lightweight, single-shot, breach-loaded, pump-action, shoulder-fired weapon.

2. What is the maximum range of the M203?
   400 meters.

3. What is the maximum effective range of the M203 against a point target?
   200 meters.

4. What is the maximum effective range of the M203 against an area target?
   350 meters.

5. If you have a misfire when using the M203, how long should you wait before opening the breech for unloading the weapon?
   Thirty seconds.

6. Identify the types of ammunition available for the M203.
   * *Practice.*
   * *High explosive.*
   * *CS chemical.*
   * *Parachute illumination.*
   * *Smoke.*

7. What is the maximum effective range of the M203 against a bunker aperture?
   50 meters.

8. What is the minimum arming range of the M203?
   14 to 38 meters.

9. What is the weight of the M320 grenade launcher?
   3.3 pounds.

10. What are the three major parts of the M320?
    Grenade launcher, day/night sight, and a laser range finder.

11. What are some advantages of the M320 over the M203?
    The M320 can be used attached to a rifle or used independently with the attachment of a stock.
    The M320 can fire a variety of longer rounds that the M203 cannot fire.

12. What is the rate of fire for the M320?
    Five to seven rounds a minute.

MK19, 40mm Grenade Machine Gun, Mod3

REFERENCE: FM 3-22.27, MK19, 40mm Grenade Machine Gun, Mod3

1. Describe the MK19.
   The MK19 is an air-cooled, blowback-operated, belt-fed machine gun that fires a 40mm grenade round.

2. What is the maximum range of the MK19?
   2,212 meters.

3. What are the rates of fire for the MK19?
   - *Sustained: 40 rounds per minute.*
   - *Rapid: 60 rounds per minute.*
   - *Cyclic: 325 to 375 rounds per minute.*

4. What is the maximum effective range of the MK19?
   1,500 meters for point targets and 2,212 meters for area targets.

5. What is the immediate action for a stoppage with the MK19?
   Recharge the weapon and attempt to fire.

6. What are the two most common malfunctions of the MK19?
   Runaway gun and sluggish action.

M240B Machine Gun

REFERENCE: FM 3-22.68, Crew Served Weapons

1. The M240B was designed for tank and armored vehicles. What makes it useable for infantry or ground soldiers?
   The installation of a kit that includes a flash suppressor, front sight, butt stock, pistol grip, rear sight assembly, feed tray cover with an optical rail, a protective heat shield for the barrel, a carrying handle, and a bipod.

2. Describe the M240B.
   The M240B is a 7.62mm, belt-fed, air-cooled, gas-operated, fully automatic machine gun that fires from the open bolt position.

3. What is the cyclic rate of fire for the M240B?
   650 to 950 rounds per minute.

4. What are the rates of fire for the M240B?
   - *Sustained rate is one hundred rounds per minute.*
   - *Rapid rate is two hundred rounds per minute.*

5. What is the maximum effective range of the M240B?
   1100 meters.

6. What is the maximum range of the M240B?
   3725 meters.

7. What is the weight of the M240B?
   27.6 pounds.

8. How is the M240B fed?
   The M240B is metal link belt fed.

9. What types of ammunition can be used with the M240B?
   - *M80-ball.*
   - *M61-armor piercing.*
   - *M62-tracer.*
   - *M63-dummy.*
   - *M82-blank.*

10. When should the M240B's barrel be changed?
    - *After every minute of continuous or cyclic fire.*
    - *After two minutes of rapid fire.*
    - *After ten minutes of sustained fire.*

11. What is the M240B's ammunition adapter?
    The adapter allows a gunner to use the one hundred-round carton and bandoleer.

12. What is the clearing process for the M240B?
    - *Move the safety to the fire "F" position.*
    - *With the right hand (palm up) pull the cocking handle to the rear and ensure the bolt is locked to the rear.*
    - *Return the cocking handle to the forward position.*
    - *Place the safety on safe "S."*
    - *Raise the cover assembly and conduct the four point safety check for brass, links, or ammo.*
    - *Check the feed pawl assembly*
    - *Check the feed tray*
    - *Lift the feed tray and inspect the chamber*
    - *Check the space between the bolt and chamber*
    - *Close the feed tray and cover assembly and place the safety in the fire "F" position.*
    - *Pull the cocking handle to the rear and pull the trigger while manually riding the bolt forward.*
    - *Close the ejection port cover.*

13. How often should the M240B be lubricated?
    - *Before firing.*
    - *After firing.*
    - *After every ninety days of inactivity or storage.*

14. When the gunner field strips the M240B what groups or assemblies will he have?
    - *Cover assembly.*
    - *Barrel assembly.*
    - *Butt stock and buffer assembly.*
    - *Receiver assembly.*
    - *Trigger housing assembly.*
    - *Drive spring and rod assembly.*
    - *Bolt and operating rod assembly.*

M249 Squad Automatic Weapon (SAW)

REFERENCE: FM 3-22.14, M249 Light Machine Gun

1. Describe the SAW.
   An air-cooled, belt- or magazine-fed, gas-operated, automatic weapon that fires from the open-bolt position.

2. Name two unique features of the SAW.
   It has a regulator to change the rate of fire, and is fed by M16 rifle magazines as well as the standard split-link belt.

3. What ranges are marked on the rear sight drum of the SAW?
   From 300 to 1,000 meters in 100-meter increments.

4. What is the weight of the SAW with a two hundred-round box of ammunition?
   22.08 pounds.

5. What are the rates of fire for the SAW?
   Sustained rate is 85 rounds per minute; rapid rate is 200 rounds per minute; and cyclic rate is 750 rounds per minute.

6. What is the maximum range of the SAW?
   3,600 meters.

7. What is the SAW's maximum extent of grazing fire obtainable over uniformly sloping terrain?
   600 meters.

8. What are the eight major groups of the SAW?
   (1) Trigger.
   (2) Gas cylinder.
   (3) Bipod.
   (4) Handguard.
   (5) Barrel.
   (6) Receiver.
   (7) Operating rod.
   (8) Buttstock and shoulder assembly group.

9. What are the three assault firing positions used with the SAW?
   Hip, shoulder, and underarm.

M9, 9mm Pistol

REFERENCES: FM 3-22.35, Combat Training with Pistols;
TM 9-1005-317-10, Operator's Manual for Pistol, Semiautomatic, 9mm, M9

1. Describe the M9, 9mm pistol.
   A 9mm, semiautomatic, double-action, recoil-operated, magazine-fed, handheld weapon.

2. Name the ammunition authorized for use in the M9 pistol.
   M882, 9mm cartridge.

3. How many rounds will the magazine for the M9 hold?
   Fifteen.

4. Describe the magazine for the M9.
   A standard, staggered box magazine designed to hold fifteen rounds.

5. What are the safety features of the M9, 9mm pistol?
   A decocking and safety lever and a firing-pin block.

6. How many major components are there to the M9 pistol?
   Five.

7. List the major groups that make up the M9.
   - *Slide assembly.*
   - *Barrel assembly.*
   - *Receiver assembly.*
   - *Recoil spring and guide.*
   - *Magazine assembly.*

8. What is the special warning with the M9, 9mm pistol?
   The weapon will fire from the half-cocked position if the trigger is pulled.

9. What is the maximum effective range of the M9 pistol?
   50 meters.

10. State the maximum range of the M9 pistol.
    1,800 meters.

11. What is the basic load for the M9 pistol?
    Forty-five rounds.

12. What is the trigger pull for the M9 pistol?
    Single action is 5.5 pounds, and 12 pounds when fired double action.

M11, 9mm Pistol

REFERENCE: FM 3-22.35, Combat Training with Pistols

1. Describe the M11 Pistol.
   The M11 is a 9mm, handheld, magazine-fed, recoil-operated, double action, semiautomatic pistol.

2. What is the basic load for the M11 pistol?
   Thirty-nine rounds.

3. What is the maximum range of the M11 pistol?
   1800 meters.

4. What is the maximum effective range of the M11 pistol?
   50 meters.

5. What safety features are on the M11 pistol?
   * *De-cocking lever.*
   * *Automatic firing pin lock.*

6. What 9mm round can be used with the M11 but not the M9 pistol?
   The 9mm subsonic jacketed hollow point.

7. What is the functioning cycle of the M11 pistol?
   * *Feeding.*
   * *Chambering.*
   * *Locking.*
   * *Firing.*
   * *Unlocking.*
   * *Extraction.*
   * *Ejecting.*
   * *Cocking.*

8. The M11 pistol is manufactured in the civilian world as what model?
   The Sig P226.

9. How many rounds can be loaded into to the M11 magazine?
   Thirteen rounds.

10. Do the M9 and M11 fire the same ammunition?
    No. Both pistols use the M882 ball and the M917 dummy rounds. Only the M11 can use the 9mm subsonic jacketed hollow point.

M2 Machine Gun, .50 Caliber, HB, Browning

REFERENCE: FM 3-22.65, Machine Gun, Browning .50 Caliber, M2

1. Describe the M2, .50 caliber machine gun.
   A link-belt-fed, recoil-operated, air-cooled, crew-served machine gun capable of single shot as well as automatic fire.

2. What is the weight of the M2 mounted on the M3 tripod?
   128 pounds.

3. What is the weight of the M2 machine gun without the tripod?
   84 pounds.

4. For what is the M2 machine gun used?
   • *Fire support for the infantry.*
   • *Defense against low-flying aircraft.*
   • *Destruction of lightly armored vehicles.*
   • *Protection in the defense.*
   • *Reconnaissance by fire of suspected enemy positions.*

5. State the maximum range of the M2 machine gun firing M2 ball.
   6,764 meters.

6. What is the maximum effective range of the M2, .50 cal?
   1,830 meters (area) and 1,500 meters (single shots).

7. What is the muzzle velocity of the .50 cal?
   3,050 feet per second.

8. What is the sustained rate of fire for the M2, .50 cal?
   Forty rounds or less per minute.

9. What is the rapid rate of fire for the M2, .50 cal?
   More than forty rounds per minute.

10. What is the cyclic rate of fire for the M2, .50 cal?
    450 to 550 rounds per minute.

11. What must be set on the M2, .50 cal after reassembly?
    Head space and timing.

Maintenance of Equipment

REFERENCE: DA PAM 750-8, The Army Maintenance
Management System (TAMMS) User Manual

1. What does the acronym TAMMS stand for?
   The Army Maintenance Management System.

2. List the five echelons of maintenance.
   (1) Operator.
   (2) Organizational.
   (3) Direct support.
   (4) General support.
   (5) Depot.

3. What is considered the cornerstone of the entire maintenance
   system?
   Operator and crew preventive maintenance.

4. What does the acronym PMCS stand for?
   Preventive Maintenance Checks and Services.

5. When is a PMCS required to be performed?
   *Before, during, and after operation of a piece of equipment; and
   at weekly and monthly intervals.*

6. Who is required to perform a PMCS?
   Every operator who is assigned a piece of equipment.

7. Name the manual used to perform a PMCS.
   The operator's manual (-10 series).

8. After completing operation of a piece of equipment, what is the
   last thing you should do before turning in the equipment?
   An after-operations PMCS.

9. List the three types of records to be maintained for a piece of
   equipment.
   Operational, maintenance, and historical.

10. List the forms that are normally carried in the equipment record
    folder for a routine dispatch.
    • *DA Form 2404.*
    • *DA Form 2408-14.*

- *DD Form 1970.*
- *SF Form 91.*
- *DD Form 518.*

11. What form is used to record repaired faults corrected by replacing parts?
    DA Form 2404.

12. What is a DA Form 2408-14 used for?
    To record uncorrected faults and deferred maintenance.

13. What is a DA Form 2408-14?
    Uncorrected fault record.

14. What type of deficiency cannot be recorded on the DA Form 2408-14?
    A deadline deficiency.

15. What is a Standard Form 91?
    Operator's Report of Motor Vehicle Accident.

16. What is a DD Form 1970?
    Motor Equipment Utilization Record.

17. Identify the regulation that covers Prevention of Motor Vehicle Accidents.
    AR 385-55.

18. What is the maximum number of hours a driver may continuously drive a military vehicle?
    Ten hours.

19. What should drivers do during driving breaks?
    Inspect their vehicles for faults and ensure that equipment and cargo are secure.

20. Why do commanders assign designated and assistant operators to military vehicles and equipment?
    To foster pride in ownership and to designate responsibility for operator maintenance.

21. At a minimum, when must chocks (blocks) be used?
When parked on an incline; when doing maintenance; and when loaded on railcars.

22. What vehicle should be placed at the rear of a convoy?
The largest nonpassenger and nonhazardous cargo-carrying vehicle in the convoy, to serve as a protective convoy block.

23. What does the acronym MAIT stand for?
Maintenance Assistance and Instruction Team.

24. What does NMC stand for?
Not mission capable.

25. In accordance with (IAW) DA Pamphlet 738-750, for what does the symbol "L" stand?
Required lubrication or lubrication performed.

26. Identify a Class I leak.
Seepage of fluid not great enough to form drops.

27. Identify a Class II leak.
Seepage of fluid great enough to form drops but not enough to cause drops to drip while the item is being inspected.

28. Identify a Class III leak.
Leakage of fluid great enough to form drops that fall while the item is being inspected.

29. What is a DA Form 5504?
Maintenance Request Form.

30. What does SAMS stand for?
Standard Army Maintenance System.

31. What is the purpose of TAMMS?
To assist commanders in managing equipment use, equipment maintenance, and repair operations to maintain equipment to Army standards.

Maps, Map Reading, and Land Navigation

REFERENCE: FM 3-25.26, Map Reading and Land Navigation

1. What accurate information can maps give when used correctly?
   - *Distances, locations, and heights.*
   - *Best routes.*
   - *Key terrain features.*
   - *Cover and concealment information.*

2. Describe the purpose of a map.
   To provide accurate information about the locations of and the distances between ground features such as terrain, structures, populated areas, routes for travel, and communications.

3. What is a map?
   A graphic representation of a portion of the earth's surface drawn to scale upon a flat plane.

4. Distinguish the three main map sizes.
   Large, medium, and small.

5. What is a large-scale map?
   A map at a scale of 1:75,000 or larger.

6. What is a small-scale map?
   A map with a scale larger than 1:600,000 but smaller than 1:75,000.

7. Name the five basic colors found on a map. What does each color represent?
   Black—man-made features; blue—water; green—vegetation; brown—contours and relief; red—main roads and special features.

8. What colors are normally used on a map overlay and what do they represent?
   Blue—friendly forces; red—enemy forces; green—engineering obstacles; yellow—contaminated areas; black—boundaries.

9. What color is used to represent large cities on a map?
   Black.

10. What is the neat line on a map?
    The border line around the edge of the map.

11. List the five major terrain features found on a map.
    Hill, ridge, valley, saddle, and depression.

12. What are the five minor terrain features found on a military map?
    Draw, spur, cliff, cut, and fill.

13. As an aid to identifying specific terrain features, what does SOSES stand for?
    Shape, orientation, size, elevation, and slope.

14. Name three ways the relief of terrain is depicted on a map.
    • *Contour lines.*
    • *Layer tinting.*
    • *Form lines.*
    • *Shaded relief.*
    • *Hachures.*

15. What is the index line on a map?
    Every fifth contour line on a map. It is heavier (darker) than the other contour lines.

16. What do contour lines on a map represent?
    High and low ground (elevation).

17. What is contour interval?
    The vertical distance between contour lines. The amount of contour interval is given in the marginal information on a map.

18. What shape are the contour lines that indicate a valley and a draw?
    U-shaped contour lines indicate a valley; V-shaped lines indicate a draw.

19. Identify the type of terrain represented when contour lines are uniform and wide apart.
    A gentle slope.

20. In what direction does the curve of a contour line point when it crosses a stream?
    It always points upstream.

21. What are parallels of latitude?
    Measured distances going north or south of the equator.

22. Define longitude.
    A measure of distance east or west of the prime meridian.

23. Meridians of longitude run north-south and converge at the poles. Through what town does the prime meridian pass?
    Greenwich, England.

24. How many norths are there on a military map? What are they?
    Three—true north, magnetic north, and grid north.

25. What is the difference between grid north and magnetic north called?
    Grid magnetic angle (GMA).

26. Describe a declination diagram.
    A diagram that shows the interrelationship between magnetic north, grid north, and true north. It enables the user to orient a map properly.

27. Where is the legend of a map located?
    The lower left margin.

28. If you found a symbol on a map that was unknown to you, what might explain it to you?
    The marginal data located on the outside lower portion of the map.

29. Using the marginal information on a map, how do you determine the straight-line distance between two points?
    By using the bar scales.

30. To locate a point to the nearest 1,000, 100, or 10 meters, what do we use?
    To 1,000 meters, a four-digit coordinate; to 100 meters, a six-digit coordinate; to 10 meters, an eight-digit coordinate.

31. Indicate the biggest difference between a standard road map and a military map.
    The standard road map does not indicate elevation and is not marked with grids.

32. What is vertical distance?
The height (elevation) of an object above or below a datum plane (usually sea level).

33. What is a bench mark?
A man-made marker showing elevation.

34. State the rule for reading military grid coordinates.
Read right and then up.

35. What is the distance between grid lines on a map?
1,000 meters.

36. What do topographical symbols represent?
Natural or man-made objects.

37. Are topographical symbols drawn to scale on a map?
No.

38. How is an unimproved dirt road shown on a military map?
A single or double dotted line.

39. What must be done to a map before it can be used?
It must be oriented.

40. Name two ways to orient a map.
By terrain association and by use of a compass.

41. Describe the fastest and most accurate way to orient oneself with a map.
Use a compass.

42. Where does the arrow on a compass always point?
Magnetic north.

43. Describe the precautions you should take when using a magnetic compass.
Do not take readings near visible metallic masses, iron, or electrical circuits, such as metal helmets, weapons, and power lines.

44. How many sights does the compass have?
Two.

45. The lensatic compass has two scales—an outer black ring and an inner red ring. Which one is used to find directions in degrees?
The inner red ring.

46. The lensatic compass has a bezel ring. When the ring is fully rotated, how many times does it click?
120 times.

47. Each bezel ring click is equal to how many degrees?
Three.

48. What unit of measure is used mainly in artillery, tank, and mortar gunnery?
The mil.

49. There are 360 degrees in a circle. How many mils are in a circle?
6,400 mils.

50. How many mils are in one degree?
17.7 mils.

51. Describe the two special features that permit use of the lensatic compass at night.
Luminous markings and a three-degree rotating bezel with a clicking device.

52. Define "azimuth."
A horizontal angle, measured in a clockwise manner from a north base line, expressing a direction.

53. Name two methods of measuring an azimuth.
Compass and protractor.

54. How do you convert a magnetic azimuth to a grid azimuth?
By subtracting the GM angle.

55. Directions are expressed as units of angular measure. What are the three common units of angular measurement?
Degree, mil, and grad.

56. A grad is a unit of angular measure. How many grads are in a circle?
    Four hundred.

57. What is a "back azimuth" and what rule determines it?
    The reverse direction of an azimuth. If an azimuth is 180 degrees or less, add 180 degrees to obtain the back azimuth. If an azimuth is more than 180 degrees, subtract 180 degrees to find the back azimuth.

58. What is the back azimuth of 190 degrees?
    10 degrees.

59. How many known locations must you find on a map and the actual ground in order to plot your location accurately?
    At least two.

60. What does the term resection mean?
    Finding your location on a map by sighting two or more known locations with a compass or straightedge. Your location is the point at which lines (back azimuths) from the known locations intersect.

61. What does the term intersection mean?
    Finding the location of an unknown point by sighting from two or more known points.

62. Indicate three elements that must be known to travel by dead reckoning.
    A known starting point, a known distance, and a known azimuth.

63. List the three field-expedient methods for determining direction.
    Shadow-tip, watch, and North Star methods.

64. For what do the initials UTM stand?
    Universal Transverse Mercator.

65. What is an aerial photomap?
    An actual picture of the earth's surface, which shows it as it appears from the air.

66. If a military map is not available, what can you use as a substitute?
    - *Foreign map.*
    - *Atlases.*
    - *Tourist road map.*
    - *City/utility map.*
    - *Field sketches.*
    - *Aerial photograph.*

67. Name four of the eight types of maps?
    (1) Planimetric map.
    (2) Topographic map.
    (3) Photo map.
    (4) Joint operations graphics.
    (5) Photomosaic.
    (6) Terrain model.
    (7) Military city map.
    (8) Special map.

68. What are the three types of desert terrain?
    Mountain, rocky plateau, and sand or dune.

69. What three things do you need to know to navigate by dead reckoning?
    (1) A known starting point.
    (2) A known distance to travel.
    (3) A known azimuth to take.

70. Contour lines that indicate a saddle on a map look like what known object?
    An hourglass.

71. What are the eight cardinal directions?
    (1) North—N.
    (2) Northeast—NE.
    (3) East—E.
    (4) Southeast—SE.
    (5) South—S.
    (6) Southwest—SW.
    (7) West—W.
    (8) Northwest—NW.

Marching and Bivouacking

REFERENCE: FM 21-18, Foot Marches

1. What are the two classifications of troop movements?
   Tactical and administrative.

2. What is a foot march?
   The movement of troops and equipment mainly by foot with limited support by vehicles.

3. Foot marches are characterized by what?
   * *Combat readiness.*
   * *Ease of control.*
   * *Adaptability to terrain.*
   * *Slow rate of movement.*
   * *Increased soldier fatigue.*

4. What characterizes a successful foot march?
   Troops arriving at their destination at the prescribed time in a state where they are physically able to execute the tactical mission.

5. List the four types of marches.
   (1) Day.
   (2) Limited visibility.
   (3) Forced.
   (4) Shuttle marches.

6. Mention factors that influence the conduct of a foot march.
   * *Location of enemy forces.*
   * *Terrain.*
   * *Weather.*
   * *Activity of enemy aviation.*

7. What factors commonly affect a foot march?
   * *Distance of the march.*
   * *Planning effectiveness.*
   * *March discipline.*
   * *Supervision.*
   * *Time available.*
   * *Physical condition of the troops.*
   * *Training status.*
   * *Soldiers' attitude.*

8. Describe the considerations when planning a march.
   * *Tactical situation.*
   * *Morale.*
   * *Weather and terrain.*
   * *Individual loads.*
   * *Water discipline.*
   * *March discipline.*
   * *Acclimatization.*

9. Define "strip map."
   A schematic diagram of a route of march that shows landmarks and checkpoints and the distances between them.

10. Name the three foot march formations.
   Route column, tactical column, and approach march.

11. Illustrate the normal formation for a tactical road march.
   A column of twos with one file on each side of the road.

12. When marching on a road during daylight, what is the normal distance maintained between individuals?
   2 to 5 meters.

13. Explain the primary duty of the pacesetter.
   To maintain the rate of march ordered by the commander.

14. Why should a pacesetter be of medium height?
   So an average stride can be taken and maintained.

15. State the normal length of march for a twenty-four-hour period.
   20 to 32 kilometers, marching four to eight hours at 4 kilometers per hour.

16. What are the normal halts or rest cycles during a foot march?
   A fifteen-minute halt after the first forty-five minutes of walking, and a ten-minute halt after every fifty minutes thereafter.

17. Specify the normal load limits carried by a soldier on a foot march.
   48 pounds for a fighting load and 72 pounds for a normal approach march load.

18. Why is it important to wear your ruck and load-bearing equipment correctly?
    Ill-fitting or incorrectly worn equipment will chafe and tire the wearer because of unequal weight distribution.

19. What is the first step to take when constructing a two-person tent from shelter halves?
    Spread the two halves on the ground, fastening them together and attaching the guy lines.

20. When constructing a shelter or tent, what things, other than a sleeping mat or air mattress, can be used for padding on the ground to insulate against the cold or dampness?
    A poncho and/or clean grass, tender branches, leaves, or pine straw.

21. What is a march in excess of 32 kilometers in a twenty-four-hour period considered to be?
    A forced march.

22. Forced foot marches are usually accomplished by what means?
    By increasing the hours marched each day rather than the rate of march.

Military Customs and Courtesies

REFERENCES: AR 600-25, Salutes, Honors, and Visits of Courtesy;
TC 3-21.10, Drill and Ceremonies

1. Define "military courtesy."
   The respect soldiers show to each other.

2. Of all the forms of military courtesy, which is considered the most important?
   The hand salute.

3. Describe the hand salute.
   A greeting exchanged between military personnel.

4. What should accompany all salutes?
   The greeting of the day.

5. What NCOs are not addressed as "Sergeant?"
   Corporals, first sergeants, and sergeants major.

6. How is a member of the Army addressed when his or her name and rank are not known?
   The member is addressed as "soldier."

7. When should you salute indoors?
   When under arms; when reporting to an officer; at formal indoor ceremonies and parades.

8. Define the term "under arms."
   Carrying a weapon by sling, holster, or other means, or the wearing of a pistol belt and headgear.

9. When a group of soldiers is at work on a detail and an officer approaches, what should happen?
   The person in charge of the detail should salute; the rest should continue to work.

10. What should a soldier do if he or she is running and encounters a situation requiring a salute?
    Come to a walk and then salute.

11. Generally, enlisted members do not exchange salutes. Name two exceptions to this rule.
    When rendering reports in formation; when reporting to an enlisted president of a board.

12. Is an officer ever required to salute an enlisted person first?
    Yes, when the enlisted person has been awarded the Medal of Honor.

13. When do individuals in a formation salute?
    Only on the command "Present arms."

14. If you are standing outdoors, in uniform, and you hear "Retreat" being played, what should you do?
    Face toward the flag or music and come to attention; on the first note of "To the Color," render the hand salute.

15. When meeting an officer in the open, how far away should you be before rendering the hand salute?
    Approximately six paces when your paths will come close; within speaking distance when making eye contact.

16. When does the driver of a motor vehicle salute?
    Only when the vehicle has stopped and the engine is not running.

17. Is a salute given in public conveyances?
    No.

18. When is an officer not required to return a salute?
    At pay call, when he is the pay officer.

19. Give four examples of occasions when you are not required to salute.
    (1) When engaged in a work detail.
    (2) When actively engaged in athletics.
    (3) When a challenge is required on guard.
    (4) When imprisoned.

20. When are you not allowed to salute because you have lost your right to do so?
    When you are a prisoner (serving time).

21. Why are military prisoners not permitted to salute?
   Because they have dishonored their profession.

22. When is the uniform hat or cap raised as a form of greeting?
   Never.

23. When should a soldier under arms remove his or her headgear?
   - *When seated as a member or in attendance at a court or board.*
   - *When entering a place of worship.*
   - *When indoors and not on duty.*
   - *When in attendance at an official reception.*

24. When walking with someone who is senior in rank to you, where should you walk?
   On the senior person's left.

25. Upon entering a vehicle, who enters first—the junior or the senior person?
   The junior enters first. Upon leaving the vehicle, the senior goes first.

26. When foreign soldiers are invited by U.S. forces to participate in a parade, where do they march in relation to the American soldiers?
   Ahead of the U.S. elements. As a special compliment, a small escort of honor, composed of U.S. soldiers, precedes the foreign soldiers.

27. Describe what should happen if a group of soldiers riding in a vehicle hears the national anthem being played.
   The individual in charge should dismount from the vehicle and salute; the remainder should sit at attention.

28. In all ceremonies, parades, or similar functions, what military service takes precedence and why?
   The Army, because it is the senior military service.

29. What ceremony takes place at 1200 hours on Independence Day?
   A salute to the nation; one gun is fired for each state.

30. What is the Army's motto?
   This we'll defend.

31. What is the title of the Army's official marching song, and when was it formally dedicated?
"The Army Song," dedicated on Veterans Day, 11 November 1956.

32. What is Tattoo and when is it played?
Tattoo means lights out and quiet in the barracks. It is played at 2100 hours.

33. Explain the original purpose of retreat.
Its purpose was to notify sentries to start challenging until sunrise and to tell the rank and file to go to their quarters and stay there.

34. What are the recommended components of a burial escort for an enlisted person?
   - *NCOIC.*
   - *Firing party.*
   - *Pallbearers.*
   - *Bugler.*

35. How many volleys are fired over the grave at a military funeral and why?
Three volleys are fired. The custom dates back to Roman times when they said farewell three times to their dead soldiers.

36. How many guns are fired in a salute to the Union?
One for each state.

37. What are some military honors to persons?
   - *Cannon salutes.*
   - *Honor guards.*
   - *Parades or review of troops.*
   - *Ruffles and flourishes.*

Military History

1. Which military system of the ancient world has had the most influence on modern military doctrine?
   Roman.

2. When and how was the Army founded?
   On 14 June 1775 six companies of expert riflemen formed in Pennsylvania, two in Maryland, and two in Virginia.

3. When did the Revolutionary War begin and formally end?
   1775 to 1783.

4. What does the "truck" on top of the flagpole represent?
   The shot heard around the world, which started the Revolutionary War.

5. When was the Declaration of Independence adopted?
   4 July 1776.

6. Who is the individual given the most credit for improving the training and efficiency of Washington's army?
   Frederick von Steuben.

7. Who was the first Army inspector general (IG)?
   Baron von Steuben.

8. In what year did the major fighting of the Revolutionary War end?
   1781, after the British surrender at Yorktown.

9. In what year was the Constitution of the United States written?
   1787.

10. What is the oldest Army Regiment?
    The Old Guard (the 3rd Infantry Regiment).

11. In what year was West Point founded?
    1802.

12. What war was fought from 1812 to 1815?
    The War of 1812.

13. What war was fought from 1861 to 1865?
    The Civil War.

14. In which war were the most American soldiers killed?
    The U.S. Civil War.

15. Where was the first battle of the Civil War fought?
    Fort Sumter.

16. Who was the commander of troops for the Confederate Army in the Civil War?
    Jefferson Davis, the Confederate president.

17. When was the U.S. Army Air Corps established?
    1909.

18. Who was the first person to be assigned a serial number?
    General Pershing.

19. Who demanded, "Send me men who can shoot and salute?"
    General Pershing, during World War I.

20. What was the title of General MacArthur's famous speech at West Point?
    "Duty, Honor, Country."

21. What is 7 December 1941 remembered for?
    The Japanese attack on Pearl Harbor.

22. Who said, "I shall return," and where did he say it?
    General Douglas MacArthur in the Philippines.

23. In what year was the Army Airborne born?
    1942.

24. The Japanese naval fleet was decisively defeated in which battle of the Pacific theater?
    Battle of Midway.

25. What happened on 6 June 1944?
    D-Day, the Allied invasion of Europe.

26. Who was the most famous general of the Third U.S. Army during World War II?
    General George S. Patton.

27. What do the six wreaths on the Tomb of the Unknown Soldier represent?
    The six major campaigns of World War II.

28. What is the inscription on the Tomb of the Unknown Soldier?
"Here rests in honored glory, an American soldier known but to God."

29. When did the Army Air Corps become the U.S. Air Force?
September 1947.

30. Name the member of the 35th Division in WWI who became president.
Harry S. Truman.

31. Who was the first Sergeant Major of the Army?
Sgt. Maj. William Wooldridge.

32. What flag, unveiled in June 1999, is scarlet and white, and has an emblem, yellow fringe, and tassels of scarlet and white?
The sergeant major of the Army flag.

33. How many five-star generals were there, and who were they?
Five—Eisenhower, Bradley, MacArthur, Marshall, and Pershing.

34. Who was the most recent five-star general in the American Army?
General Douglas MacArthur.

35. Name four officers who held the rank of general of the Army.
Douglas MacArthur, Omar Bradley, George C. Marshall, Dwight D. Eisenhower.

36. Who is Lt. Col. Nancy J. Currie?
An Army female astronaut.

37. Who was the first African-American West Point graduate?
1st Lt. Henry O. Flipper.

38. Who was Benjamin Oliver Davis Sr.?
The nation's first African-American brigadier general.

39. Which president signed an executive order that ended racial discrimination in the military?
Harry S. Truman.

40. During the Korean War, who was the overall commander of the United Nations forces?
General Douglas MacArthur.

41. What was Gen. Douglas MacArthur's rank when he retired?
    General of the Army.

42. What president signed the Code of Conduct?
    President Eisenhower, on 17 August 1955.

43. When was the U.S. Army flag approved?
    By executive order 10607, 12 June 1956.

44. When was the position of Sergeant Major of the Army established?
    4 July 1966.

45. What American spy ship was captured by North Korean forces in 1968?
    USS Pueblo.

46. What Army officer was implicated in a massacre of civilians at My Lai during the Vietnam conflict?
    Lt. William Calley.

47. Who commanded U.S. forces in Operation Desert Storm (Iraq, 1991)?
    General Norman Schwartzkopf.

48. Who commanded U.S. forces in Operation Enduring Freedom (Afghanistan, 2001) and Operation Iraqi Freedom (2003)?
    General Tommy Franks.

49. In what year were the majority of American troops pulled out of Vietnam?
    1973.

50. When was the NCO support channel recognized as a formal entity of the Army?
    20 December 1976.

51. What happened in April 1865?
    General Lee surrendered to General Grant at Appomattox Court House. President Abraham Lincoln was assassinated.

Military Justice

REFERENCES: AR 27-10, Military Justice;
FM 27-14, Legal Guide for the Soldier

1. What is the foundation for military law?
   The Constitution.

2. When was the Uniform Code of Military Justice (UCMJ) enacted?
   1951.

3. What did the UCMJ replace?
   In 1951 the UCMJ replaced the Articles of War, which had been in existence in various forms since 1775.

4. Who makes or writes the laws contained in the Uniform Code of Military Justice?
   Congress.

5. What is the highest military court?
   The Court of Military Appeals.

6. Who makes up the Court of Military Appeals?
   Three judges appointed by the president of the United States. They are the final authority on matters of law in court-martial cases.

7. What are the purposes of nonjudicial punishment under Article 15?
   To correct and reform soldiers, to preserve an offender's record of service, and to dispose of minor infractions in a manner requiring less time and personnel than a trial by court-martial.

8. What type of offenses are normally disposed of by nonjudicial punishment?
   Minor offenses.

9. Who may administer punishment under Article 15?
   Any commanding officer or a warrant officer if in command.

10. Are there any circumstances under which an NCO is authorized to impose nonjudicial punishment?
    No.

11. Is an NCO authorized to order an enlisted soldier into arrest or confinement?
Yes.

12. What is the one circumstance in which a soldier does not have the right to refuse an Article 15 and demand trial by court-martial instead?
When the soldier is aboard ship.

13. What is the maximum number of days a soldier may wait before registering an appeal concerning punishment under an Article 15?
Five.

14. When given a Summarized Article 15, what rights does a soldier have?
Rights to a hearing, to call witnesses, and to appeal.

15. What punishments can be given under a Summarized Article 15?
- *Fourteen days' restriction, fourteen days of extra duty.*
- *Oral admonishment.*
- *Oral reprimand.*

16. If given a Summarized Article 15, does the soldier have the right to consult a lawyer?
No.

17. Can a company commander reduce a person who is in grade E-5 or above?
No.

18. Can NCOs in the pay grades E-7 through E-9 be reduced under Article 15 action?
No.

19. When a soldier is reduced under Article 15, what is the effective date of reduction?
The date the commander imposed punishment.

20. What is the maximum punishment a company commander may impose under Article 15?
- *Oral admonishment or reprimand.*
- *Fourteen days' restriction.*

- *Fourteen days of extra duty.*
- *Loss of seven days' pay.*
- *Reduction in rank.*

21. Can extra duty include duty similar to the conduct of a personal servant?
    No.

22. Who may act on an appeal of an Article 15?
    The next commander in the normal chain of command.

23. Can an officer be administered an Article 15?
    Yes.

24. Is a reprimand an authorized punishment under Article 15?
    Yes.

25. When pay is forfeited by an individual under Article 15, what is done with the money?
    It is contributed to the Soldiers' Home Fund (old soldiers' home).

26. For how long can an Article 15 be displayed on the unit bulletin board?
    Seven days.

27. What are the guidelines for placing an NCO on extra duty?
    The extra duty cannot degrade the NCO's rank.

28. What are the two types of Article 15 proceedings?
    Summarized and Formal.

29. Imposed extra duty must meet what guidelines?
    The extra duty must not be of a cruel or unusual nature. It cannot be duty that is normally considered an honor to perform. It can't cause ridicule, be a safety or health hazard, or demean the soldier's position as a sergeant or specialist.

30. What is the maximum punishment a field grade officer may impose under Article 15 proceedings?
    - *Admonishment or reprimand.*
    - *Forty-five days' extra duty.*

- *Sixty days' restriction.*
- *Half of one month's pay for two months.*
- *Loss of one or more pay grades for E-4 and below.*
- *Loss of one pay grade for E-5 or E-6.*
- *Four days' confinement with a restricted diet when aboard ship.*

31. What article in the Uniform Code of Military Justice (UCMJ) prohibits self-incrimination?
Article 31.

32. What is an Article 32 hearing?
A formal hearing, much like a grand jury investigation in the civilian community, to determine if there is enough evidence to bring charges before a court-martial.

33. Which article in the Manual for Courts-Martial gives the NCO his authority?
Article 91.

34. State the three types of courts-martial.
Summary, special, and general.

35. Mention the minimum number of members for each type of court-martial.
Summary—none; special—three; and general—five.

36. Can a soldier demand trial by court-martial in lieu of punishment under UCMJ?
Yes, except for Army members embarked in vessels.

37. What does Article 138 of the UCMJ do?
It provides soldiers with a means of redress against their commanding officer if they feel they have been wronged.

38. What time limit for filing charges is established under Article 138 of the UCMJ?
A soldier must act within ninety days of learning of a wrong.

Miscellaneous Questions

1. For what does the acronym MOS stand?
   Military occupational specialty.

2. Is your sergeant major in your chain of command?
   No, he or she is in the NCO support channel.

3. What is DD Form 2A?
   U.S. Armed Forces Identification Card.

4. What does the fourth character in your primary military occupational specialty (PMOS) represent?
   Your skill level.

5. What are some of the requirements for promotion to E-5 and E-6?
   • *Meet time-in-grade and time-in-service requirements.*
   • *Be in a promotable status.*
   • *Be recommended by the unit commander.*
   • *Appear before a selection board and be recommended.*
   • *Be fully qualified in the MOS in which recommended.*
   • *Meet education requirements.*

6. What is a DA Form 6?
   A duty roster form.

7. What is the purpose of the NCOES system?
   To prepare selected soldiers for performing duties appropriate to their grade or next higher grade; to provide training in appropriate supervisory skills; to increase general skills, confidence, and a sense of pride and esprit de corps.

8. Which rank in the Army provides its basic strength?
   Private.

9. Where does the chain of command stop?
   The U.S. president going up and the private going down.

10. Among soldiers of the same rank, how is precedence determined?
    By date of rank, by time of active service, and by age.

11. Can a chaplain assume command when no other officer is present?
No, but chaplains have authority to exercise functions of operational supervision and control.

12. What is the NCO support channel?
A channel of communication among NCO leaders that parallels and reinforces the chain of command.

13. To whom does the Sergeant Major of the Army serve as an enlisted advisor?
The Chief of Staff of the Army.

14. How much authority is delegated to each level in the chain of command?
Only as much as the leader needs to accomplish assigned duties and responsibilities.

15. For what does the acronym DEERS stand?
Defense Enrollment Eligibility Reporting System.

16. What is the purpose of DEERS?
To verify enrolled family members as well as their eligibility to receive military benefits and privileges.

17. What is the maximum number of promotion points that can be awarded by a promotion board?
150

18. For what does the acronym ACES stand?
Army Continuing Education System.

19. For what does the acronym SHORAD stand?
Short-range air defense weapons.

20. For what does the acronym MANPAD stand?
Man-portable air defense weapons.

21. Name two man-portable air defense systems.
The Redeye and the Stinger.

22. What is the purpose of DA Form 1155?
   Casualty reporting.

23. List the four casualty statuses that can be entered on a DA Form 1155.
   (1) KIA (dead).
   (2) WIA (wounded).
   (3) MIA (missing).
   (4) Captured.

24. What is a battle drill?
   A collective action executed by a platoon or smaller unit without applying a deliberate decision-making process.

25. What is the first sentence of the NCO creed?
   "No one is more professional than I."

26. There is a saying in regards to the color scheme on live Army ordnance. Do you know what it is?
   Green and yellow can kill a fellow.

27. When are enlisted soldiers required to have an initial official DA Photograph taken?
   • *On promotion to SSG.*
   • *On appointment to CSM.*

28. When are enlisted official DA Photographs required to be updated?
   • *Every fifth year after the initial.*
   • *After receiving an award of ARCOM or above.*
   • *After any major change in physical and or uniform appearance.*

29. Can your official DA Photograph be altered?
   No. There is a zero tolerance policy on altering photographs.

30. What is the Army Post Deployment Health Reassessment Program?
   It is a program designed to identify soldiers with physical and behavioral health concerns after returning from a combat area.

31. What is Leaders' Net?
   Leaders' Net is an online program at AKO to enhance collaboration, communications, and knowledge sharing to advance the study and application of Army Leadership and Leader Development.

Nuclear, Biological, and Chemical Warfare (NBC)

REFERENCES: FM 3-11.4, NBC Protection; FM 3-11.5,
NBC Decontamination

1. What are the three types of nuclear bursts?
   Air, ground, and subsurface.

2. What are the three effects of a nuclear explosion?
   Blast, heat, and radiation.

3. What three types of rays are emitted by an atomic explosion?
   Alpha, beta, and gamma rays.

4. Which of the three effects of a nuclear explosion does the most damage?
   The blast, because of flying objects and high winds.

5. What should you do during and after a nuclear attack?
   Take cover, decontaminate yourself, and continue the mission.

6. What is an IM 93/UD?
   A type of radioactivity dosimeter for personnel.

7. State three factors that influence the effectiveness of biological agents.
   Light, temperature, and moisture.

8. What are the four types of microorganisms found in biological agents?
   (1) Bacteria.
   (2) Fungi.
   (3) Rickettsiae.
   (4) Viruses.

9. Where would a biological attack be most effective?
   - *Cities.*
   - *Large troop concentrations.*
   - *Animals.*
   - *Vegetable crops.*

10. What is the best protection against biological agents?
    Personal hygiene and immunization shots.

11. Name the four classes or types of chemical agents.
    (1) Nerve.
    (2) Blister.
    (3) Blood.
    (4) Choking.

12. How may chemical agents be deployed?
    - *Aerial spray.*
    - *Artillery.*
    - *Bombs.*
    - *Individuals (pollution of water, food, supplies).*

13. What type of agent smells like newly mown hay or green corn?
    Choking agent.

14. Is tear gas (CS) a toxic agent?
    No.

15. What substance, used to detect liquid chemical agents, gives a red, yellow, or green signature?
    M-8 chemical-agent detection paper.

16. What are the procedures for using a latrine in a chemical environment?
    - *Decon your gloves.*
    - *Pull up your protective overgarment enough to open the protective trousers.*
    - *Take off your protective gloves.*
    - *Pull down or adjust BDU pants.*
    - *Use the latrine.*
    - *Decon any skin that may have become contaminated.*
    - *Pull up or adjust BDU pants.*
    - *Put gloves back on.*
    - *Pull up or adjust protective trousers.*
    - *Pull down overgarment coat and fasten snaps in the back.*

17. What does MOPP stand for?
    Mission-oriented protective posture.

18. How many levels of MOPP are there?
    Five.

19. Specify the five levels of MOPP.
    Zero—Protective equipment carried.
    One—Overgarments (trousers and coat) worn, the rest carried.
    Two—Overgarments and boots worn, the rest carried.
    Three—Overgarments, boots, and mask worn.
    Four—Overgarments, boots, mask, and gloves worn with all zippers closed and closures fastened.

20. What are the colors of chemical, biological, and radiological contamination markers?
    Chemical—yellow with red letters (GAS); biological—blue with red letters (BIO); radiological—white with black letters (ATOM).

21. At the first indication of an NBC attack, what should you do?
    Stop breathing and mask.

22. Once a unit has masked, who can make the decision to unmask?
    The unit commander.

23. When a detector kit is not available, what procedures should you use to unmask?
    (1) Have two or three soldiers hold their breath and break the seals of their masks for fifteen seconds while keeping their eyes open.
    (2) Have those soldiers reseal, clear, check their masks, and wait ten minutes.
    (3) If no symptoms appear, have the same soldiers break the seals on their masks and take two or three deep breaths.
    (4) Have them again reseal, clear, check their masks, and wait ten more minutes.
    (5) If no symptoms appear, have the same soldiers unmask.
    (6) Wait for ten minutes.
    (7) If no symptoms appear, have all soldiers unmask, but continue to be aware of any symptoms that may appear.

24. In what direction does the face of the protective mask point when packed into the carrying case?
    Outward.

25. How often should you check your protective mask during peace-time conditions?
    Every six months or after every test or training session.

26. List the three methods of decontamination.
    Removing, neutralizing, or destroying the agent.

27. When crossing a contaminated area, what two things should you
    try to avoid doing?
    Stirring up dust and touching anything.

28. What are the two standard decontaminants in use in today's
    Army?
    *Supertropical bleach (STB) and decontaminating solution #2
    (DS-2).*

29. If you mix STB and DS-2, what will the results be?
    Fire (spontaneous combustion).

30. What are the four basic uses of smoke on the battlefield?
    Obscuring, screening, protecting, and marking.

31. What are some biological attack indicators?
    • *Mysterious illness.*
    • *Large number of insects or unusual insects.*
    • *Large number of dead wild or domestic animals.*
    • *Artillery bursts with low powered explosives.*
    • *Aerial bombs that pop rather than explode.*
    • *Mist or fog sprayed by aircraft.*

32. What is some expedient cover that you can use in a nuclear
    attack?
    • *Beneath tracked vehicles.*
    • *Walls.*
    • *Ditches.*
    • *Culverts.*
    • *Basements of buildings.*

33. What are the three types of immediate decontamination?
    Skin decon, personal wipe down, and operator's spray down.

34. What are the three levels of decontamination?
    Immediate, operational, and thorough.

35. What are the problems a person must deal with when in full MOPP?
    - *Heat stress.*
    - *Hunger.*
    - *Thirst.*
    - *Performance degradation.*
    - *Reduced ability to see and hear.*

36. What would make individual protective equipment unserviceable?
    - *If it is cracked, ripped, or torn.*
    - *If a fastener is broken or missing.*
    - *If petroleum products are spilled or splashed on it.*
    - *If rubber parts become sticky.*

37. What three types of NBC warning are there?
    Verbal, metal-on-metal banging, and hand and arm signals.

38. What warning should you heed when using decontamination wipes or powders?
    Do not wipe your eyes, mouth, or open wounds. Use water to flush these areas.

39. What do you do with the autoinjector after injecting?
    Pin it prominently on the jacket of the person receiving the injection.

40. Where would you attach M9 paper to your MOPP gear?
    If right handed, on your right upper arm, left wrist, and right ankle.

41. What could cause the M9 paper to give a false positive?
    Hot, dirty, oily, or greasy surfaces.

Physical Readiness Training

REFERENCE: FM 21-20, Physical Fitness Training;
TC 3-21.20, Army Physical REadiness Training

1. Name the three basic types of fitness programs.
   Unit, individual, and special.

2. What are the commands to get a unit from a normal line formation into an extended rectangular formation?
   - *Extend to the left, March.*
   - *Arms downward, Move.*
   - *Left, Face.*
   - *Extend to the left, March.*
   - *Arms downward, Move.*
   - *Right, Face.*
   - *From front to rear, Count off.*
   - *Even numbers to the left, Uncover.*

3. What is the command to return the extended rectangular formation back to the original formation?
   Assemble to the right, March.

4. What formation is used for guerrilla exercises?
   Circle.

5. When forming the extended rectangular formation, what is the command to uncover?
   Even numbers to the left, Uncover.

6. When forming the extended rectangular formation, the first command, "Extend to the left," is given. What should the next command be?
   Arms downward, Move.

7. When you take the Army physical fitness test, what is the minimum number of points you must score in an event to pass?
   Sixty points.

8. What are the two standards or categories for the Army physical fitness test (APFT)?
   Initial entry training standards and Army standards.

9. What are the three phases of physical conditioning?
   Initial conditioning, toughening, and sustaining.

10. Identify the principles of physical training.
    • *Progression.*
    • *Overload.*
    • *Balance.*
    • *Variety.*
    • *Regularity.*
    • *Specificity.*
    • *Recovery.*

11. What is the objective of physical fitness training?
    To develop soldiers who are physically capable and ready to perform their duties and missions in peace or during combat.

12. According to FM 3-2, what is considered satisfactory weight loss?
    $1/2$ to $1 1/2$ pounds a week.

13. Cite the components of physical fitness.
    • *Cardiorespiratory endurance.*
    • *Muscular strength.*
    • *Muscular endurance.*
    • *Flexibility.*
    • *Agility.*
    • *Coordination.*
    • *Body composition.*

14. What is the most efficient way for soldiers to build aerobic endurance?
    By running.

15. Name two alternate aerobic exercises.
    Swimming, bicycling, cross-country skiing.

16. How is flexibility developed and maintained?
    By stretching exercises.

17. What are the components of "motor fitness?"
    • *Speed.*
    • *Agility.*
    • *Muscle power.*

- *Hand-eye coordination.*
- *Foot-eye coordination.*

18. In strength training, when you exercise to a point where you have the inability to perform another repetition, what is it called?
Working to failure.

19. What type of exercise places the maximum workload on the muscle throughout the entire range of motion?
Isokinetic.

20. What wastes physical training time?
- *Unprepared and unorganized leaders.*
- *Too large a group.*
- *Insufficient training intensity.*
- *Rates of progress that are too slow or fast.*
- *Extreme formality.*
- *Inadequate facilities.*
- *Rest periods that are too long.*

21. In planning physical fitness training, what items need to be addressed for safety reasons?
- *Environmental conditions.*
- *Soldiers' levels of condition.*
- *Facilities.*
- *Traffic.*
- *Emergency procedures.*

22. List three groups of individuals that may require a special physical training program.
(1) Individuals who fail the APFT and do not have a profile.
(2) Individuals who are overweight.
(3) Individuals who have permanent or temporary medical profiles.
(4) Individuals who are over forty and have not been medically screened.

23. What do the initials FITT stand for?
F—Frequency.
I—Intensity.
T—Time.
T—Type.

24. What are the three types of fitness programs?
Unit, individual, and special.

25. All exercise sessions should include what?
Stretching during the warm up and the cool down.

26. What is considered critical to the success of a good physical training program?
Effective leadership.

27. What are the components of good physical fitness programs?
- *Weight control.*
- *Diet.*
- *Nutrition.*
- *Stress management.*
- *Spiritual and ethical fitness.*

28. What is physical readiness?
The ability to meet the physical demands of any combat or duty position, accomplish the mission, and continue to fight and win.

29. What are the three principles followed during the conduct of physical readiness training?
Precision, progression, and integration.

30. What are the three components of physical readiness training?
Mobility, strength, and endurance.

31. What are the physical readiness performance factors?
Agility, balance, coordination, flexibility, posture, stability, speed, and power.

32. The physical readiness training system has what three types of training?
Off-ground, on-ground, and combatives.

Religious Support

REFERENCE: FM 1-05, Religious Support

1. What is the primary military agency for ensuring the free exercise of religion for soldiers?
The Chaplain Corps.

2. Who is responsible for ensuring the religious freedom of soldiers?
Army Commanders.

3. What support do Army Chaplains provide?
  • *Religious services.*
  • *Rites.*
  • *Sacraments.*
  • *Ordinances.*
  • *Pastoral care.*
  • *Religious education.*
  • *Family Life ministry.*
  • *Institutional ministry.*
  • *Support to Command and Staff.*
  • *Religious and humanitarian support.*

4. What are the three religious support functions?
(1) Nurture the living.
(2) Care for the dying.
(3) Honor the dead.

5. Can an Army Chaplain carry a weapon or side arm?
No. The Chaplain is a noncombatant and will not bear arms in accordance with AR 165-1.

Security and Intelligence

REFERENCE: AR 380-5, DA Information Security Program Regulation

1. Can a person with a secret clearance be denied access to any secret material?
   Yes, access is granted on a need-to-know basis.

2. What does the acronym SALUTE stand for?
   —Size.
   —Activity.
   —Location.
   —Unit.
   —Time.
   —Equipment.

3. What is DA Form 672 used for?
   Safe or Cabinet Security Record; it is a record of the access and locking of containers.

4. What are the three security classifications used to protect material against disclosure?
   Top Secret, Secret, and Confidential.

5. List the five Ss that apply to POWs.
   • *Search.*
   • *Segregate.*
   • *Silence.*
   • *Speed to the rear.*
   • *Safeguard.*

6. What is SAEDA?
   Subversion and espionage directed against the U.S. Army.

7. What would make an individual a possible target for SAEDA approach?
   Drug or alcohol abuse, indebtedness, or any situation or incident for which an individual could be bribed or blackmailed.

8. How many overall security classifications can a document have?
   One.

9. Who is authorized access to classified information?
   A person with a need to know, with a clearance equal to the protected information.

10. How is top secret material destroyed?
    By burning or by pulping.

11. Into what groups should POWs be divided?
    • *Officers.*
    • *NCOs.*
    • *Privates.*
    • *Deserters.*
    • *Females.*
    • *Civilians.*
    • *Political personnel.*

12. Name some counterintelligence measures.
    • *Camouflage.*
    • *Use of sign and countersign.*
    • *Recon.*
    • *Censorship.*
    • *Noise and light discipline.*
    • *Operation security (OPSEC) training.*

13. Who is responsible for security?
    Everyone.

14. Can items such as cash and jewelry be stored with classified material?
    No.

15. Who has the responsibility for reporting lost or compromised classified material?
    Anyone having knowledge.

16. What two markings will not be used to identify classified material?
    "For official use only" and "Limited official use."

17. Are persons granted access to classified materials by virtue of rank?
    No.

18. What color is the secret label on security containers?
    Red.

19. What color is the confidential label on security containers?
    Blue.

20. What color is the unclassified label on security containers?
    Green.

21. What color is the top secret label on security containers?
    Orange.

21. Explain operations security.
    The process of identifying critical information and analyzing friendly actions pertaining to military operations to prevent information and knowledge of actions getting to the enemy.

Supply Economy, Clothing, and Equipment

1. Define the term "supply economy."
   Obtaining the maximum use of supplies and equipment by the use of correct materials and equipment for a specific job and by preventing waste.

2. Who is responsible for supply accountability?
   Every soldier.

3. List some supply economy practices.
   • *Turn off water when done.*
   • *Turn out lights when they are not needed.*
   • *Maintain equipment.*
   • *Don't use government property for personal use.*

4. What is a common supply economy practice used in dining facilities?
   Portion control.

5. How many classes of supplies are there?
   Ten.

6. What are expendable supplies?
   Any supplies that are consumed through use, such as food, paper, paint, pens, or pencils.

7. State the first thing you should do before signing a hand receipt.
   Inventory the items for accountability and inspect for serviceability.

8. When you sign a hand receipt, what does your signature establish?
   Direct responsibility for the article or equipment.

9. Name four ways you can be relieved from accountability for property.
   • *Turn in.*
   • *Report of survey.*
   • *Statement of charges.*
   • *Cash collection voucher.*

10. Describe a report of survey.
    A survey conducted by a responsible party to determine whether the loss, damage, or destruction to government property was caused by neglect.

11. When initiating a report of survey, what are the time constraints?
    Within fifteen calendar days after the discovery of the discrepancy.

12. How many days does it take to initiate a report of survey?
    Five days.

13. What is a statement of charges?
    A statement signed by an individual allowing the payroll deduction of charges for government property that was lost, damaged, or destroyed through neglect.

14. What form is used for a statement of charges?
    DD Form 362.

15. What should a soldier do with personal and military issue items before going on leave if not planning on packing them?
    Inventory the items with his or her squad leader and turn them in to supply for safekeeping.

16. In reference to military clothing, what does "initial issue" mean?
    The complete issue of all personal military clothing a soldier receives when first entering the Army.

17. Before signing your clothing record, what column should you check?
    The issue column above the place for your signature.

18. What is referred to as the "gig line" on the uniform?
    The alignment of the shirt, belt buckle, and trouser fly.

19. Name three items of military clothing that can be worn with civilian clothing.
    Low quarters, the all-weather coat, and the windbreaker.

20. Specify the purpose of the Army clothing allowance.
    To replace initial issue items that become unserviceable.

21. How long must a soldier be in the Army before he or she can receive a clothing allowance?
At least six months.

22. What facility on post offers a soldier payroll-deduction laundry service?
Quartermaster laundry.

23. Explain the general purposes of inspections.
To maintain soldiers' health and welfare, and to ensure standards for accountability and serviceability of government property.

24. Name three types of inspections.
Inspection in ranks, full field inspection, and showdown inspection.

25. What sort of things does an inspecting officer look for during an inspection?
- *Uniformity of displays.*
- *Cleanliness of soldiers and equipment.*
- *Accountability of property.*
- *Serviceability of property.*

26. What is property accountability?
The obligation of a person to keep an accurate formal record of property issued to him or her.

27. What is property responsibility?
The obligation of a person to ensure that government property entrusted to his or her possession is property used and cared for.

Survival

REFERENCES: FM 21-76-1, Multiservice Procedures for Survival, Evasion, and Recovery; FM 3-21.75, Combat Skills of the Soldier

1. Name four crucial factors in staying healthy.
   - *Having adequate food and water.*
   - *Practicing good personal hygiene.*
   - *Getting sufficient rest.*
   - *Keeping your immunizations up-to-date.*

2. Why should you drink the milk only from green coconuts rather than from mature or ripe coconuts?
   The milk from mature coconuts contains an oil that acts like a laxative.

3. If you are forced to eat wild plants, what should you avoid?
   - *Plants that have a milky sap or a sap that turns black when exposed to the air.*
   - *Plants that are mushroom-like.*
   - *Plants that resemble onion, garlic, parsley, parsnip, or dill.*
   - *Plants that have carrot-like leaves, roots, or tubers.*

4. Mention several methods of improving the taste of wild plants.
   - *Soaking.*
   - *Parboiling.*
   - *Cooking.*
   - *Leaching.*

5. Specify the difficulties in operating in arid areas.
   - *It is hard to find food, water, and shelter.*
   - *Physical movement is demanding.*
   - *Land navigation is difficult.*
   - *There is limited cover and concealment.*

6. List the environmental factors you must consider when operating in an arid area.
   - *Low rainfall.*
   - *Intense sunlight and heat.*
   - *Wide temperature ranges.*
   - *Sparse vegetation.*

- *High mineral content of ground surfaces.*
- *Sandstorms.*
- *Mirages.*
- *Light levels.*

7. What heat effects must be considered in an arid area?
   - *Direct sun.*
   - *Reflective heat gain.*
   - *Conductive heat gain.*
   - *Hot, blowing winds.*

8. State the basic rules for water consumption in an arid area.
   At temperatures below 100 degrees Fahrenheit, drink one pint of water every hour. At temperatures above 100 degrees Fahrenheit, drink one quart of water every hour.

9. Name five of the nine common signs of dehydration.
   (1) Dark urine with a very strong odor.
   (2) Low urine output.
   (3) Dark, sunken eyes.
   (4) Fatigue.
   (5) Loss of skin elasticity.
   (6) Emotional instability.
   (7) Thirst.
   (8) Trench line down center of tongue.
   (9) Delayed capillary refill in the fingernail beds.

10. In a survival situation, what are some of the emotions you are likely to experience?
    - *Fear.*
    - *Anxiety.*
    - *Anger.*
    - *Frustration.*
    - *Depression.*
    - *Loneliness.*
    - *Boredom.*
    - *Guilt.*

11. Each letter in the word survival stands for a survival concept; for what does each letter stand?
    S—Size up the situation.
    U—Use all your senses.
    R—Remember where you are.
    V—Vanquish fear and panic.
    I—Improvise.
    V—Value living.
    A—Act like the natives.
    L—Live by your wits, but for now, learn basic skills.

12. In the area of evasion and survival, what do the letters BLISS mean?
    B—Blend.
    L—Low silhouette.
    I—Irregular shape.
    S—Small.
    S—Secluded location.

13. What does the pattern signal "V" stand for?
    Require assistance.

14. What does the pattern signal "X" stand for?
    Require medical assistance.

15. In a survival environment, what can the common cattail provide?
    It can provide a food source and has medical use for wounds, sores, inflammation, and burns.

16. What is the most important thing to have in a survival situation?
    The will to survive.

Uniforms and Insignia

REFERENCE: AR 670-1, Wear and Appearance of
Army Uniforms and Insignia

1. When must identification tags be worn?
   • *When engaged in field training.*
   • *When traveling in aircraft.*
   • *When outside Continental United States (CONUS).*
   • *When directed by the commander.*

2. When are you not required to wear headgear while in uniform?
   Indoors (unless under arms or directed by the commander); when wearing headgear would interfere with the safe operation of military vehicles or equipment; or while in a privately owned or commercial vehicle.

3. Where are unit awards worn on the uniform?
   Over the right breast pocket.

4. Can the Army green uniform be used for formal social occasions?
   Yes, when worn with a black bow tie and a white shirt.

5. What Army regulation covers the appearance and wear of uniforms?
   AR 670-1.

6. How are overseas service bars placed on the Army green uniform?
   Centered on the outside half of the right sleeve, 4 inches above and parallel to the bottom of the sleeve.

7. How are service stripes placed on the Army green uniform?
   4 inches up from the bottom of the left sleeve at a 45-degree angle, with the lower end toward the inside seam of the sleeve.

8. How far below the pocket seam are marksmanship badges placed?
   $^1/_8$ inch below the seam.

9. How are service ribbons worn?
   In order of precedence from wearer's right to left, in one or more rows, centered $^1/_8$ inch above the left breast pocket, with either no spaces or $^1/_8$-inch spaces between rows.

10. What is the distance between badges on the class A green uniform?
    1 inch.

11. Describe the wear of the battle dress uniform (BDU) cap.
    The cap is to be worn straight on the head so that the cap band creates a straight line around the head and is parallel to the ground.

12. When is the wearing of the black four-in-hand tie required with
the Army Blue Uniform?
When the uniform is worn before retreat.

13. What is the proper length of the skirt when worn with the female
class A uniform?
*No more than 1 inch above or 2 inches below the crease in the
back of the knee.*

14. Identify the jewelry that may be worn while in uniform.
* *Watch.*
* *ID bracelet.*
* *No more than two rings.*
* *Tie clasp with a tie.*
* *Religious medallion on a chain, provided it is concealed.*

15. What are the dimensions of the dress uniform nameplate?
1 inch × 3 inch × 1/10 inch.

16. What is the rule for the placement of enlisted brass on the uniform?
Uncle Sam is always right.

17. List the four categories of badges that are worn on the Army
uniform.
(1) Marksmanship badges and tabs.
(2) Combat and special skill badges and tabs.
(3) Identification badges.
(4) Foreign badges.

18. With which uniforms is it all right to expose pens in the pockets?
*Hospital duty, food service, combat vehicle crewman, flight, and
the ACU coat.*

19. How is the beret worn?
*It is worn with the headband straight across the forehead and 1
inch above the eyebrows. The flash is positioned above the left eye
and the excess material is draped over the right ear.*

20. When can a wireless Bluetooth device or non-wireless earpiece be
worn by soldiers in uniform?
*Only while operating a commercial (POV) or military vehicle
(includes motorcycles and bicycles).*

21. Can combat, special skill, and identification badges be worn on
the ACU in the field or in deployed environments?
No.

22. What caps may be worn with the Improved Physcial Fitness Uniform?
    Black knit cap, foliage green microfleece cap, black microfleece cap.

23. What socks can be worn with the ACU?
    Black, tan, or green socks.

24. What is the Army policy on tattoos?
    Tattoos or brands visible above the Class A uniform color are prohibited; tattoos or brands that are extremist, indecent, sexist, or racist are prohibited anywhere on the body.

25. What is the composition of the Improved Physical Fitness Uniform?
    - *Jacket: running, gray and black.*
    - *Pants: running, black.*
    - *Trunks: running, gray and black.*
    - *T-shirt: gray, short sleeve, and gray, long sleeve.*
    - *Cap: knit, black.*

26. Why was the Blue Army Uniform selected as the standard Army Service Uniform?
    The Blue Army Uniform dates back to 1779 and General Washington.

27. What is not authorized for wear with the Blue Army Service Uniform when worn after retreat?
    Combat boots and organizational items such as Brassards, MP accessories, and distinctive unit insignia.

28. When will wear of the Blue Army Service Uniform become mandatory?
    The 4th Quarter of FY 2014.

29. What is the mandatory wear out date for the Green Army Service Uniform?
    The 4th Quarter of FY 2014.

Appendix A

Army Regulations and Field Manuals

To assist you in improving your knowledge of military subjects, a list of materials and resources is provided here.

A very wise sergeant major once told me that the most knowledgeable solders were not those who could quote sections of regulations, but rather those who knew where to look to find needed information. I have found this statement to be true. The Army changes; regulations and field manuals change. If a decision is important, or if you want to be on top of everything, check. Don't rely on your memory.

The following is a list of resources that soldiers and NCOs need to be knowledgeable about in order to do well in their day-to-day activities, not only to prepare for local boards but also to manage careers and be able to train and take care of their soldiers. The Army regulations and field manuals listed are for the "average" soldier. Because of your MOS you may want to list additional publications. While it isn't necessary for you to have your own complete library of military publications, it is important that you know where to find them. Your local Personnel and Administration Center (PAC) and learning resource centers will have copies for your use.

Army Regulations (AR)

27-10	Military Justice
220-45	Duty Rosters
350-30	Code of Conduct, Survival, Evasion, Resistance, and Escape (SERE) Training
380-5	DA Information Security Program Regulation
600-8-8	Army Sponsorship Program
600-20	Army Command Policy
600-8-22	Military Awards
600-9	The Army Weight-Control Program
600-25	Salutes, Honors, and Visits of Courtesy
600-85	Army Substance Abuse Program
608-1	Army Community Service
621-5	Army Continuing Education System
623-3	Evaluation Reporting System
640-30	Photographs for Military Personnel Files
670-1	Wear and Appearance of Army Uniforms and Insignia
840-10	Flags, Guidons, Streamers, Tabards, and Automobile and Aircraft Plates
930-4	Army Emergency Relief
930-5	American Red Cross

Field Manuals (FM)

1-05	Religious Support
3-05.30	Psychological Operations
3-05.70	Survival
3-11.4	NBC Protection
3-11.5	NBC Decontamination
3-22-20	Army Physical Readiness Training
3-21.75	The Warrior Ethos and Soldier Combat Skills
3-22.9	M16A1, M16A2, M4 Rifle Marksmanship
3-22.27	MK19, 40mm Grenade Machine Gun
3-22.65	Machine Gun, Browning .50 Caliber, M2
3-22.68	Crew Served Machine Guns, 5.56mm and 7.62mm (M60, M240B, M249 SAW)
3-23.25	Shoulder-Launched Munitions
3-23.30	Grenades and Pyrotechnic Signals
3-24	Counter Insurgency
3-25.26	Map Reading and Land Navigation
3-34.214	Explosives and Demolitions

4-25.11	First Aid
4-25.12	Unit Field Sanitation Team
5-34	Engineer Field Data
6-22	Military Leadership
6-22.5	Combat Stress
7-0	Training Units and Developing Leaders for Full-Spectrum Operations
7-1	Battle Focused Training
7-22.7	The NCO Guide
7-21.13	The Soldier's Guide
8-34	Food Sanitation for the Supervisor
20-3	Camouflage, Concealment, and Decoys
21-10	Field Hygiene and Sanitation
21-18	Foot Marches
21-20	Physical Fitness Training
21-75	Combat Skills of the Soldier
23-23	Antipersonnel Mine M18 (Claymore)
27-10	The Law of Land Warfare

Training Circulars

| 3-21.10 | Drill and Ceremonies |
| 3-22.20 | Army Physical Readiness Training |

Army Doctrine Publications

| 7-0 | Training Units and Developing Leaders |

Training Manuals

| 3-23.31 | 40mm Grenade Launcher |
| 3-23.25 | Shoulder-Launched Munitions |

Pamphlets (PAM)

DA 350-58	Leader Development for America's Army
TRADOC 600-4	IET Soldier's Handbook
DA 750-8	Users Manual for The Army Maintenance Management System (TAMMS)

Appendix B

Chain of Command

Listed here are the personnel normally found in the formal chain of command within an Army unit. Check with your first sergeant to fill in the missing names and be sure to keep the list up-to-date. In some Army units, the terminology for the positions within the chain is not the same. Also, the length of the chain or the number of personnel within a unit's chain is not always the same. The basic chain is represented here:

Commander in Chief   President _____

Secretary of Defense   Honorable _____

Secretary of the Army   Honorable _____

Chairman of the Joint
    Chiefs of Staff   _____

Chief of Staff of the Army General _____

CG, Major Army Command   General _____

*Corps Commander   _____

*Division/Command Commander _____

*Brigade/Group Commander   _____

Battalion Commander   _____

Company Commander   _____

Platoon Leader   _____

Squad/Section Leader   _____

The NCO support channel complements and parallels the chain of command and provides a structure for the day-to-day activities of the Army. This channel is outlined here:

Sergeant Major of the Army   _____

Major Army Command CSM   _____

*Corps CSM   _____

*Division/Command CSM   _____

*Brigade/Group CSM   _____

Battalion CSM   _____

First Sergeant   _____

Platoon Sergeant   _____

Squad/Section Leader   _____

Note: An * indicates titles may vary depending on the type of command, e.g., support commands, service support commands, etc.

Appendix C

Code of Conduct for Members of the U.S. Armed Forces[*]

I
I am an American, fighting in the forces which guard my country and our way of life. I am prepared to give my life in their defense.

II
I will never surrender of my own free will. If in command, I will never surrender the members of my command while they still have the means to resist.

III
If I am captured I will continue to resist by all means available. I will make every effort to escape and aid others to escape. I will accept neither parole nor special favors from the enemy.

IV
If I become a prisoner of war, I will keep faith with my fellow prisoners. I will give no information or take part in any action which might be harmful to my comrades. If I am senior, I will take command. If not, I will obey the lawful orders of those appointed over me and will back them up in every way.

---

*As amended, May 1988.

V

When questioned, should I become a prisoner of war, I am required to give my name, rank, service number, and date of birth. I will evade answering further questions to the utmost of my ability. I will make no oral or written statements disloyal to my country and its allies or harmful to their cause.

VI

I will never forget that I am an American, fighting for freedom, responsible for my actions, and dedicated to the principles which made my country free. I will trust in my God and in the United States of America.

Soldiers should be aware that our government also has responsibilities as a result of the Code of Conduct. When a soldier is at war, and especially if captured by the enemy and imprisoned, the government will do the following:

- Keep faith with you and stand by you as you fight in its defense;
- Care for your family and dependents; and
- Use every practical means to contact, support, and gain release for you and all other prisoners of war.

Appendix D

The NCO Creed

No one is more professional than I. I am a Noncommissioned Officer, a leader of soldiers. As a Noncommissioned officer, I realize that I am a member of a time-honored corps, which is known as "The Backbone of the Army." I am proud of the Corps of Noncommissioned Officers and will at all times conduct myself so as to bring credit upon the Corps, the Military Service, and my country regardless of the situation in which I find myself. I will not use my grade or position to attain pleasure, profit, or personal safety.

Competence is my watchword. My two responsibilities will always be uppermost in my mind—accomplishment of my mission and the welfare of my soldiers. I will strive to remain tactically and technically proficient. I am aware of my role as a Noncommissioned Officer. I will fulfill my responsibilities inherent in that role. All soldiers are entitled to outstanding leadership.

I will provide that leadership. I know my soldiers and I will always place their needs above my own. I will communicate consistently with my soldiers and never leave them uninformed. I will be fair and impartial when recommending both rewards and punishment.

Officers of my unit will have maximum time to accomplish their duties; they will not have to accomplish mine. I will earn their respect and confidence as well as that of my soldiers. I will be loyal to those with whom I serve; seniors, peers, and subordinates alike. I will exercise initiative by taking appropriate action in the absence of orders. I will not compromise my integrity, nor my moral courage. I will not forget, nor will I allow my comrades to forget that we are professionals, Noncommissioned Officers, leaders!

Appendix E

The Soldier's Creed

I am an American Soldier. I am a member of the United States Army—a protector of the greatest nation on earth. Because I am proud of the uniform I wear, I will always act in ways creditable to the military service and the nation it is sworn to guard.

I am proud of my own organization. I will do all I can to make it the finest unit of the Army. I will be loyal to those under whom I serve. I will do my full part to carry out orders and instructions given my unit or me.

As a soldier, I realize that I am a member of a time-honored profession—that I am doing my share to keep alive the principles of freedom for which my country stands. No matter what situation I am in, I will never do anything, for pleasure, profit, or personal safety, which will disgrace my uniform, my unit, or my country. I will use every means I have, even beyond the line of duty, to restrain my Army comrades from actions disgraceful to themselves and the uniform.

I am proud of my country and its flag. I will try to make the people of this nation proud of the service I represent, for I am an American Soldier.

Appendix F

Standing Orders, Rogers's Rangers

These orders were written in 1759 by Robert Rogers, commander of colonial rangers in the French and Indian War. The equipment has changed, but the concepts are just as valid today.

1. Don't forget nothing.

2. Have your musket clean as a whistle, hatchet scoured, 60 rounds powder and ball, and be ready to march at a minute's warning.

3. When you're on the march, act the way you would if you was sneaking up on a deer. See the enemy first.

4. Tell the truth about what you see and what you do. There is an army depending on us for correct information. You can lie all you please when you tell other folks about the rangers. But don't never lie to a ranger or officer.

5. Don't ever take a chance when you don't have to.

6. When we're on the march we march single file, far enough apart so one shot can't go through two men.

7. If we strike swamps, or soft ground, we spread out abreast, so it's hard to track us.

8. When we march, we keep moving till dark, so as to give the enemy the least possible chance at us.

9. When we camp, half the party stays awake while the other half sleeps.

10. If we take prisoners, we keep 'em separated till we have had time to examine them, so they can't cook up a story between 'em.

11. Don't ever march home the same way. Take a different route so you won't be ambushed.

12. No matter whether we travel in big parties or little ones, each party has to keep a scout 20 yards ahead, 20 yards on each flank, and 20 yards in the rear, so the main body can't be surprised and wiped out.

13. Every night you'll be told where to meet if surrounded by a superior force.

14. Don't sit down to eat without posting sentries.

15. Don't sleep beyond dawn. Dawn's when French and Indians attack.

16. Don't cross a river by a regular ford.

17. If somebody's trailing you, make a circle, come back onto your own tracks, and ambush the folks that aim to ambush you.

18. Don't stand up when the enemy's coming against you. Kneel down, lie down, hide behind a tree.

19. Let the enemy come till he's almost close enough to touch. Then let him have it and jump out and finish him up with your hatchet.

STACKPOLE BOOKS

Military Professional Reference Library

Air Force Officer's Guide
Airman's Guide
Guide to Personal Financial Planning for the Armed Forces
Army Dictionary and Desk Reference
Army Officer's Guide
Career Progression Guide for Soldiers
Combat Leader's Field Guide
Combat Service Support Guide
Enlisted Soldier's Guide
Guide to Effective Military Writing
Job Search: Marketing Your Military Experience
Military Money Guide
NCO Guide
Servicemember's Guide to a College Degree
Servicemember's Legal Guide
Soldier's Study Guide
Today's Military Wife
Veteran's Guide to Benefits

1-800-732-3669 • www.stackpolebooks.com.